"Jason Alba's I'm on LinkedIn—Now What??? *is a must read for all social-networking 'non-believers.' Not only is it a useful tool for the those looking to further their existing knowledge on how to optimize the many benefits of being 'Linked In,' it also serves as a treasure map to those who have yet to discover LinkedIn's hidden benefits and potential."*
Michelle A. Riklan, Managing Director, Riklan Resources, http://www.riklanresources.com, **and SelfGrowth.com,** http://www.selfgrowth.com

"As a career services professional, I recommend that every job seeker or professional currently working have a LinkedIn profile. Jason Alba's book is the perfect guide to help understand the important components of LinkedIn and how having an effective profile can pave the way to job opportunities. He does a great job explaining in logical, clear terms what the advantages of this site are and how candidates can come across more competitively as subject matter experts in their fields."
Dawn Rasmussen, President, Pathfinder Writing and Career Services, http://www.pathfindercareers.com

"Jason Alba has created an easy-to-follow guide for people of all skill levels. Anyone who wants to get more out of LinkedIn will find this book an invaluable resource. Jason's practical tips make LinkedIn's professional networking tool a breeze to set up and to keep up in today's marketplace. I highly recommend this."
Denny Stockdale, Author, *Conversations from the Neighborhood Ice Cream Shop: 8 Keys to Rediscovering Lost Dreams and Finding your Life's Calling*; **and President, Stockdale Resource Group,** http://www.dennystockdale.com

"LinkedIn is a very powerful tool, and Jason Alba does a tremendous job of helping a new user get the most out of it."
Carl Chapman, Founder, CEC Search, http://www.cecsearch.com

"As a new 'learner' of LinkedIn, I was overwhelmed with the dynamics of it: how it works, what to put into my bio, etc. After reading Jason Alba's book, I not only felt at least like a semi-expert, but I have referred every single client in the world to Jason's fine information. It's simply a must read for getting started and recognizing the value of the site. Good work, Jason."
Valerie Sokolosky, President, Valerie & Company,
http://www.valerieandcompany.com

"For both novices and skilled users alike, Jason's book offers a rare perspective on the value of LinkedIn for career management, while using a conversational tone that puts his readers at ease. I continue to recommend I'm on LinkedIn—Now What??? to job seekers, not only because they need a way to drill beyond the surface of the site, but also due to the little-known (but powerful) techniques he shares for business search and self-promotion. Jason's explanations of the strategy and reasoning behind LinkedIn as a personal PR and career transition tool are worth the price of admission alone!"
Laura Smith-Proulx, Executive Resume Writer, Certified Career Management Coach & Owner, An Expert Resume,
http://www.anexpertresume.com/

"Ask any ten LinkedIn users how to use LinkedIn and, I bet you, at least nine of them will tell you to speak to Jason Alba! In this third edition, Jason wastes absolutely no time with fluff—it's not his style, anyway—and delivers a power-packed book that will turn even the most confused, non-LinkedIn user into a guru. Want to learn the secret to building your profile? It's in here. Not sure how to use Groups? Look no further. Bottom line: in this book, both the newbie AND the expert LinkedIn user are well served!"
Laura M. Labovich, Chief Career and Networking Strategist, Aspire! Empower! Career Strategy Group,
http://aspire-empower.com/

THIRD EDITION

I'm on LinkedIn—Now What???

A Guide to Getting the Most OUT of LinkedIn

By Jason Alba

Foreword by Bob Burg
Author of *Endless Referrals* and
Coauthor of *The Go-Giver*

20660 Stevens Creek Blvd., Suite 210
Cupertino, CA 95014

Published by Happy About®
20660 Stevens Creek Blvd., Suite 210, Cupertino, CA 95014
http://happyabout.com

First Printing: September 2007
Second Printing: December 2008
Third Printing: March 2011
Paperback ISBN: 978-1-60005-197-5 (1-60005-197-9)
eBook ISBN: 978-1-60005-198-2 (1-60005-198-7)
Place of Publication: Silicon Valley, California, USA
Paperback Library of Congress Number: 2011920531

Dedication

To Kaisie, Samantha, William, Taylor, Kimberly, and Daniel.

I can't tell you how much I appreciate the love and support you've given me since this journey began.

Acknowledgments

Thank you to the LinkedIn team, who continues to change the product, making a new edition so necessary! When I sent someone at LinkedIn a previous manuscript, I got an email back saying that they already had major changes planned for the next few months. That's either quite frustrating, or it's my job security.

I got considerable strength and encouragement from many friends and acquaintances I've met since I got laid off five years ago. To the thousands of people who have purchased this book in a previous edition, thank you. Your comments on my blog (http://www.imonlinkedinnowwhat.com), your emails, and the evangelizing of my book have encouraged me to keep this work current.

I have a close circle of associates and partners who have been instrumental in giving me feedback, advice, and tips to help me hone my message and delivery and make sure that what I present is just what their professional, executive, and corporate clientele needs.

My publisher deserves a big hat-tip, as he encouraged me, challenged me, and believed in me since we first met in 2007. Thank you, Mitchell and Happy About, for the opportunity to change my life and career path by becoming one of your authors.

Finally, deep gratitude and respect goes to my family, immediate and extended, as I have seen them support my entrepreneurial dream, wondering what exciting thing is coming up next while allowing me to enjoy the road I'm on.

A Message from Happy About®

Thank you for your purchase of this Happy About book. It is available online at http://www.happyabout.com/linkedinhelp.php or at other online and physical bookstores.

- Please contact us for quantity discounts at sales@happyabout.info
- If you want to be informed by email of upcoming Happy About® books, please email bookupdate@happyabout.info

Happy About is interested in you if you are an author who would like to submit a non-fiction book proposal or a corporation that would like to have a book written for you. Please contact us by email editorial@happyabout.info or phone (1-408-257-3000).

Other Happy About books available include:

- Storytelling About Your Brand Online & Offline:
  http://www.happyabout.com/storytelling.php
- I'm at a Networking Event—Now What???:
  http://www.happyabout.com/networking-event.php
- I'm in a Job Search—Now What???:
  http://www.happyabout.com/jobsearchnowwhat.php
- I Need a Killer Press Release—Now What???:
  http://www.happyabout.com/killer-press-release.php
- I've Got a Domain Name—Now What???:
  http://www.happyabout.com/ivegotadomainname.php
- I've Landed My Dream Job—Now What???:
  http://www.happyabout.com/mydreamjob.php
- iPad Means Business:
  http://www.happyabout.com/ipadmeansbusiness.php
- Social Media Success!:
  http://www.happyabout.com/social-media-success.php
- Internet Your Way to a New Job:
  http://www.happyabout.com/InternetYourWaytoaNewJob.php
- Community 101:
  http://www.happyabout.com/community101.php
- 18 Rules of Community Engagement:
  http://www.happyabout.com/community-engagement.php
- 42 Rules of Social Media for Small Business:
  http://www.happyabout.com/42rules/social-media-business.php

Contents

Figures

Foreword by Bob Burg

Several years ago, I began receiving requests from friends inviting me to join a new online membership site called LinkedIn. Soon I was getting emails from people I barely knew, asking me to "join their LinkedIn network." Although I'd been using the Internet for networking and relationship building for some time, I really wasn't all that interested in LinkedIn. Still, not wanting to hurt these people's feelings, I accepted.

I went through the process of posting my profile, but didn't do much with the site. As time went on, more and more people sent me invites.

Every so often, I'd be asked for help connecting someone with someone else, who apparently knew someone who knew someone I knew. From time to time, people from specific groups who'd read my book, *Endless Referrals*[1] would notice I was a LinkedIn member and ask me to contribute an article for their newsletter. I guess you could say I'd become a part of the LinkedIn community. But I wasn't really utilizing LinkedIn in any active or significant way.

Why not? No perceived need, and no desire. (You might remember those two reasons from Sales Training 101 as the two most common reasons prospects say "no.") And why didn't I have any perceived need or desire? Because I had frankly no idea what to do or how to make LinkedIn a positive experience for me.

---

1. Bob Burg, *Endless Referrals*, McGraw Hill, 2005

That's exactly where this book comes into play. Jason Alba has done a first-rate job of solving that challenge for me, and he will for you, too.

A former unemployed IT professional and business strategist, Jason found that finding a good job, even in a "job seeker's market," was a pretty daunting task. Today, he runs a career management company, called JibberJobber. In *I'm on LinkedIn—Now What???*, he presents us with an actual system to tap into the power of the LinkedIn service.

And that's the key word here: system. That's what I was lacking in my early LinkedIn experiences.

Why is having a system so important? I define a system as "the process of predictably achieving a goal based on a logical and specific set of how-to principles." In other words, if it's been proven that by doing A you'll achieve B, then you know that all you need to do is follow A and you'll eventually achieve B. As Michael Gerber points out in his classic, The *E-Myth Revisited*[2] (slightly paraphrased): systems permit ordinary people to achieve extraordinary results—predictably.

Whatever the B is you want to achieve here, *I'm on LinkedIn—Now What???* provides you with the A for getting it. After an excellent introduction explaining exactly what LinkedIn is and how it (basically) works, Jason then walks you through a guided tour of clear principles and powerful strategies for getting the most out of your LinkedIn experience.

---

2. Michael Gerber, *The E-Myth Revisited*, HarperCollins, 1995

While Jason sees LinkedIn as an excellent business-building tool, he also looks at the site with a carefully critical eye. Jason himself began achieving great success utilizing LinkedIn only after floundering with it his first few months, and he does not hold back in pointing out its weaknesses and suggesting areas where LinkedIn could improve and make its service more valuable for its members. I found especially refreshing those passages where he points out the areas of LinkedIn where he has still not grasped its highest use. Someone that humble, I tend to trust.

Jason tells us that LinkedIn is not a replacement for your networking efforts (online or offline); rather, it is an excellent tool for facilitating some facets of your networking strategy. I absolutely concur. The creed of my Endless Referrals System®[3] is that "all things being equal, people will do business with, and refer business to, those people they know, like, and trust." And no computer or online medium is going to replace that personal connection—but it certainly can enhance it and provide potential networking contacts with more opportunities to connect. In this book, you'll learn how to do exactly that, whether it's for direct business, resources you need, helpful information, finding joint venture partners, hiring a new employee, or getting hired for a new job.

Something I particularly appreciate about Jason's approach is that he shows us not only how we can gain value from LinkedIn, but, just as important, how we can utilize LinkedIn to provide value to others. As any true networker knows,

---

3. http://www.burg.com

this is not only immensely satisfying in its own right, but it is also the best way to receive even more value oneself.

As you travel through this excellent guide, be prepared to learn from a man who has done his homework. Jason has learned what he knows the hard way, through trial and error, both his own and many other people's, and put it all between the covers of a book so that you and I can learn it all the easy way!

Best wishes for great success.

Bob Burg
Author of *Endless Referrals* and Coauthor of *The Go-Giver*

Part I
Getting Started

Part I is all about getting started on LinkedIn.

- Chapter 1 talks about my personal experience with getting started, and why I decided to write this book.

- Chapter 2 talks about what LinkedIn is and isn't, helping you understand its capabilities and limitations.

- Chapter 3 is a crash course on your profile, and what you need to do to set it up correctly, increasing your chances of being found by others and leaving a favorable impression on those who find your profile.

- Chapter 4 talks about setting up your Account & Settings so your interaction with the website reflects your needs (and so you don't feel like you are getting spammed from LinkedIn).

1 Introduction

When I revised the first edition, I found, much to my surprise, over five hundred changes that I needed to make. In this revision, I know there are many, many changes to the last edition. LinkedIn has been busy changing the interface, functionality of the site, how users interact and navigate, and much more. Indeed, I feel like LinkedIn is almost a different system than it was even a year ago.

Also, I've seen considerable changes in methods and tactics for those in job search mode and those interested in business marketing.

Social networking and social marketing, and corporate branding and personal branding have evolved as online tools. They have come and gone; they add new features and take away some features (perhaps they are "simplyfing" the navigation).

Even my own understanding of networking, relationships, marketing, and branding have evolved over time. I went from a job seeker who no one would talk to, to a new startup-CEO, to an author, to a speaker, to a seasoned startup-CEO. My goals and my relationships are different. I follow up better, I am more strategic, and I have an acute awareness of how overloaded professionals are.

If I had time, I'd do a new edition of this book at least every six months. I can't do that, but I have something better for you. If you are interested in personal career management, whether you are in a job search or you are preparing for a job search, check out the JibberJobber blog, which I update almost every day: http://www.jibberjobber.com/blog. If you are interested in LinkedIn stuff, whether for your personal career or for a company, check out my LinkedIn blog: http://imonlinkedinnowwhat.com. If you need a video tutorial of how to use LinkedIn, check out http://linkedinforjobseekers.com, where I have my three-hour LinkedIn tutorial DVD.

I was first introduced to LinkedIn in February 2006 when I was walking out of a networking meeting with a guy who came to tell us he had just landed a new job. As we were walking out, he recommended I create an account on LinkedIn. I didn't want to get an account on yet another website, and I doubted one more site would add value to my job search.

Regardless, I did get an account on LinkedIn. I found it to be a very lonely place. For the first few months, I had only six connections. I tried searching for important contacts, based on company name or title. At the time, there were about eight million users in LinkedIn (now there are over eighty million), but my searches produced no results. I thought the search function was broken! I didn't find anyone I was interested in contacting. My search for "manager" or "CEO" in "Salt Lake" produced nothing! What was wrong?

LinkedIn just wasn't working for me. I knew if I could just figure out what all the buzz was about and learn how professionals were using LinkedIn to improve their businesses and careers, I could benefit. But I didn't understand what I needed to do.

And so I began to learn what LinkedIn was all about—how to use it, how to benefit from it, and what its limits were. Once I began to understand how it was intended to be used, it became much more valuable to me in my personal career management, and later as I developed my own business.

While learning the hows and whys of LinkedIn, I found a lot of people were still confused by the same things that had confused me. It was obvious *why* you would use LinkedIn if you were a recruiter (the first book[4] on LinkedIn was written specifically for recruiters) or a job seeker. But the "why" wasn't the problem—it was the "how"! Even recruiters, who should have been power users, were confused on how to use LinkedIn!

As a business owner, I've used LinkedIn to bring considerable value to my business. I've used it to find the right contacts as I work on business development and grow my network strategically. I've done company research as I prepared to make a business proposal or do competitive intelligence research. I've used LinkedIn to enhance my personal and my corporate brands. I've positioned myself as an expert and gotten in front of other LinkedIn users. LinkedIn has become an indispensable tool for me in my overall social marketing strategy.

I hope this book can serve as a reference for you to get the best out of LinkedIn. *I'm on LinkedIn—Now What???* is not a comprehensive book on networking, nor is it a general book on social networking. There are some great books on those subjects. We'll talk about networking and social networking, borrowing from the experts and using examples from my contacts. By the time you finish this book, you should have a solid understanding of what LinkedIn is, how to use it, and why things on LinkedIn work the way they do.

A quick word about my online strategy: LinkedIn is NOT the only social tool I use. I use Facebook, Twitter, blogs, Yahoo! Groups, and other tools to help grow my network, nurture relationships, and share my brand. I use these tools as part of a complementary strategy, connecting and participating as appropriate, and then leaving the social space so I can do my day job. I have gotten a lot of value out of my multi-tool strategy. I would not stop using any of these tools and try to replace them with just one tool—they are all complementary tools that make up my comprehensive social strategy.

---

4. Bill Vick and Des Walsh, *Happy About LinkedIn for Recruiting*, Happy About, 2006

I hope this book will teach you how to use LinkedIn more effectively. I hope you will implement some of the tactics to find new business, customers, employers, vendors, partners, employees, or even friends. While there are no guarantees, I know that many people have had their professional and personal lives enriched because of LinkedIn, and I hope you can, too!

Now, let's get started!

Chapter Summary

- This book came about because of my own journey with LinkedIn.

- LinkedIn is not the only online networking tool you should use, but it is definitely one of the most powerful.

- How you get value out of LinkedIn might be different than how I get value out of LinkedIn, but many of the ideas and techniques shared here could help you get more value out of LinkedIn.

What Is LinkedIn?

Over the years, I've known people who thought LinkedIn was something that it really wasn't. It's critical to understand the boundaries of LinkedIn: what LinkedIn is and what it isn't. Depending on who you are and what you need, it may be extremely valuable, moderately valuable, or only slightly valuable. You might spend a lot of time on LinkedIn, with a proactive strategy, or hardly any time on LinkedIn, with a just-be-there or reactive strategy.

So just what is LinkedIn? Some call LinkedIn a social network, others call it a business network (explicitly stating that it is not a social network), and others call it a contact management system. I agree that it has elements of each, but I don't call it any of these things.

TIP: LinkedIn is a **tool**. Understanding the value proposition of LinkedIn, and its secondary benefits, should help you get the most out of your time or money investment in the tool.

LinkedIn is a tool to help you find others. Perhaps you should connect with the people you find so they can add value to you (or so you can add value to them).

Here are some simple definitions of the degrees of connection:

First-degree contact: someone who has agreed to connect with you, or who has sent you an invitation to connect, and you've accepted.

Second-degree contact: someone who is a first-degree contact of your first-degree contact; so, if you connect with me, all of your other contacts become my second-degree contacts.

Third-degree contact: first-degree contacts of your second-degree contacts; this is where LinkedIn gets really powerful!

Maybe you aren't sure what that value is, but it seems to make sense to connect. LinkedIn is a tool to help you expand your network, both wider (with first-degree contacts) and deeper (with contacts of your first-degree contacts).

In addition, LinkedIn is a place to be found. You should set up and optimize your own profile so others can find you. Your profile might have similarities to your resume. For example, you can list where you went to school, where you worked, dates of employment, what your tasks/roles were, and what your interests are. You can state what your current interests are (open to job opportunities, interested in hooking up with classmates, etc.), and you can choose what information to make available to other LinkedIn members and what to make available to anyone (even the general public who aren't members of, or logged into, LinkedIn). We'll talk about this later, in the chapter on "Your Profile."

The value of LinkedIn grows as more people join. Imagine there were only one thousand people in LinkedIn, and they all lived in one city and followed one profession. Would LinkedIn be valuable to you? Not unless you happened to live in that city, or had an interest in that profession. Fortunately, there are over eighty million people who have created accounts on LinkedIn, which means you have a huge database of prospective contacts. The value doesn't stop at finding new contacts.

Because of the number of users and their diverse backgrounds and interests, LinkedIn has become a place where an immense amount of information gets shared. One of the strong commonalities between LinkedIn users is a desire to network either for business or professional reasons. LinkedIn is not a site for music bands to advertise (like Myspace is), although musicians do have profiles and can use techniques in this book to promote their businesses. LinkedIn is not a site to look for dating opportunities, although I'm sure people have hooked up and perhaps even married as a result of relationships forged through LinkedIn. LinkedIn is a site where people come to develop relationships that can be professionally beneficial.

This means the users are the source of knowledge about business and political issues, how-to's, career management, job leads, consulting opportunities, and more. LinkedIn does not (yet) have all of the social networking features you see on other sites. However, it provides considerable value beyond just having network contacts.

What LinkedIn Is Not

Historically, LinkedIn has not been a social network rich with fun little communities, video and joke sharing, and other such features. But since the last edition of this book, LinkedIn has added a number of social features that brings it much closer to a Facebook or Twitter-like network.

On LinkedIn you connect with others (who become your first-degree connections), see your connections' contacts (who are your second- or third-degree contacts), and possibly interact with them through a LinkedIn forum. In my last edition, I wrote:

> With regard to social networking features, LinkedIn is not as feature-rich as the other sites, as it doesn't allow users to share blog posts, or leave messages and comments on your contacts' profiles in a conversational way for others to see.

This is no longer true. It's been said that LinkedIn is adding features and trying to become more like Facebook (while Facebook seems to be adding features to become more like Twitter!). Now, you can leave a "status update" that your network can see and comment on. You can have interesting conversations in LinkedIn Groups and bring your network together for a discussion on LinkedIn Answers. It is definitely more social than it ever was.

LinkedIn is not a contact manager in the traditional sense of contact management, although they have been adding some contact management features (I'm highly biased here because my company provides a more traditional contact management system). Again, in the last edition of this book, I said:

> A contact manager allows me to put contact information in a system and manage data for each contact. For example, I would put the name of a network contact and then update phone numbers, special dates (birthdays, graduations, etc.), spouse's name, and kids' names and ages.

Now, you can go into your contacts list and add all kinds of information, including phone numbers, tags, notes, and more. However, there are a few reasons I would not recommend you use LinkedIn as your relationship management tool:

1. It is not uncommon for LinkedIn to deny access to your account for a variety of reasons (see their User Agreement for a list of possible violations). I have received messages from people asking me how to get access back to their account. Nobody should get denied access to their relationship management tool, with notes, etc. of their contacts.
2. You can only make these annotations on a contact if they are a first-degree contact. Imagine you and I meet, we have a good discussion, and you want to nurture the relationship. You cannot make any notes on LinkedIn until we become first-degree contacts. What if I am not on LinkedIn? Or if I don't connect with you?
3. What if a first-degree contact decides to disconnect? What happens to any special data you put on their record? (I don't know.)

These are the most significant reasons I can't recommend LinkedIn as a relationship management tool. Don't get me wrong: LinkedIn is a powerful relationship tool, but it's not a relationship management tool.

You can tell I think it's a haphazard practice to use LinkedIn as your contact management system when you have so little control over who is actually in your network, what data is collected and managed, and who can see your contacts. At a minimum, calling LinkedIn a contact management tool is a misnomer. If you use LinkedIn as a relationship management tool, make sure you **back up your contacts regularly**. On my JibberJobber blog, I wrote[5] about a career coach who mistakenly had her account disabled, wondering if she would ever get her network back. The account was eventually reinstated, but I think she'll regularly back up her contacts, just in case.

Clearly, LinkedIn is a networking tool. It is not, however, a networking **silver bullet**. Timeless networking principles such as "givers gain," etiquette, long-term relationship nurturing, and investing time and effort in others are critical. LinkedIn is not a *replacement* for your networking efforts (online or offline); rather, it is an excellent tool to complement your networking strategy.

Finally, LinkedIn is not a time hog. Once you get certain things set up (mostly your profile and preferences) you don't have to worry about spending much time in LinkedIn. Of course, if you have the time, you can derive significant value by incorporating some of the more advanced features we talk about in this book.

Why Do People Use LinkedIn?

With over eighty million people in LinkedIn (there were about thirteen million sign-ups when the first edition was printed and twenty-eight million sign-ups when the second edition went to print!), you are sure to find people who use different strategies and techniques and have different motives for using LinkedIn. While it boils down to "people meeting people," here are some examples of why people sign up for, or use, LinkedIn:

- **Professionals**—to develop their personal brand; to search for potential clients; to determine job titles and positions of prospects; to research potential contacts, companies, and industries

---

5. http://tinyurl.com/linkedin-maintenance
   jibberjobber.com/blog/2008/07/17/linkedin-maintenance-
   do-this-right-now-or-else/

- **Job seekers**—to get new network contacts; to find new leads and opportunities; to network into a company (do a search on a company and see who in their network might have an "in" where they want to interview); to establish a presence and hopefully be found by recruiters, hiring managers, and HR

- **Recruiters and hiring managers**—to find prospects for open positions; to develop a rich network of prospective candidates; to search through connections of their connections, digging deeper into second- and third-degree contacts

- **Entrepreneurs**—to develop an online presence; to establish a strong brand; to meet other entrepreneurs or potential business partners, customers, and investors; to build a team of cofounders and employees; to do market research; to get publicity

There's a nice list at Web Worker Daily[6] that lists "20 Ways to Use LinkedIn Productively." What are your objectives with LinkedIn? Listing your own objectives and goals can be helpful to you as you figure out what to spend time on.

Some of the tactics presented in this book may be appropriate for you, while others may trigger ideas of how to get more out of LinkedIn. Remember, while LinkedIn may be useful to help you do your job, there is always a chance others are looking for you too, so be sure to have your profile as updated as possible. In other words, just because you are using LinkedIn to find others, don't forget that others might be evaluating you as a service provider, new talent, or for some other type of relationship.

LinkedIn Benefits

"I'm on LinkedIn—now what???" That's the question I have heard for the last four-plus years! Once you sign up for an account, it isn't obvious where all the value comes from, even though LinkedIn has a lot of evangelists. A common complaint I hear is based on the idea that people think they are signing up for the "premier professional networking site" and find their experience does not match up to all the

---

6. http://tinyurl.com/ProductiveUse
   gigaom.com/collaboration/20-ways-to-use-linkedin-productively/

hype. I think this is more an issue of education and training (what is LinkedIn and how do I use it?) than technology. That's what this book is all about.

I have strong feelings about social networking—I personally feel "social networked out." I regularly receive email invitations to social networks and wonder why I'm being invited. Sometimes the social networks are geography-based, sometimes they are interest-based. Usually the invitations don't appeal to me because I don't live in that particular geographic area, or I don't have much passion for that particular interest! Many invitations come without the inviter knowing he sent the invitation at all. For example, someone joins a new network just to see what it is all about and sometime during the sign-up process he puts his address book in, which is used to invite all of his contacts to the network as well. Some people call this spam. Unfortunately, the person who joined the network doesn't realize all of their contacts got invitations on their behalf!

Putting aside the general nuisances of social networking (some are more annoying than others), I've found it's definitely worth my time to participate actively in LinkedIn. I think about how my job search in 2006 would have been different had I been able to develop a LinkedIn network the size I have now (over three thousand connections), and how my career management, and even job performance, could have been better if I had a real LinkedIn strategy. Here are some of the benefits I have seen from my active participation in LinkedIn:

1. **Ability to be known.** Using LinkedIn, sharing my knowledge about LinkedIn on various Yahoo! and Google groups, commenting on LinkedIn blogs, and participating in LinkedIn Answers has made me much more visible. When I talk about "personal branding," I am talking about being branded as a subject-matter expert and/or a thought leader. The ways I participate online help define my personal brand. I contribute, give, and share in a positive way to develop a good reputation in various online communities.
2. **Ability to be found.** I have strategically developed my LinkedIn profile to increase the chances of showing up in search results. What search results? I've put in key phrases so people who want to hire me can find me. Who is looking to "hire" you? If someone is looking for a project manager in Seattle, will they find you? If someone is looking for a certain kind of service provider, will they

find you? There are plenty of people looking for relationships that make sense on LinkedIn—update your profile so when people look for you they can find you!

3. **Ability to find others.** If you have a network that is "big enough," which LinkedIn indicates is around sixty-five contacts, you should have sufficient reach when you do searches. I'm amazed when I do searches and come up with certain results. For example, finding someone with the last name of "Jason" who has "Oracle" in their profile. I was only able to find this person through LinkedIn because I have a network that is "big enough." If I only had five first-degree contacts, I would not have found Mr. Jason who has Oracle in his profile. I have also found new, relevant contacts through my interactions on LinkedIn Groups and LinkedIn Answers.

4. **Opportunity to learn and share.** LinkedIn Answers is an excellent tool and one of my favorite features in LinkedIn. Answers might be the only reason some users log into LinkedIn! Some people have found new business, while others have received expert advice and information faster than any other option available to them. I have used Answers to share my personal and corporate brand, share news about products (for example, when I ask for input on a new project I'm working on), etc.

5. **Ability to connect with group members.** There are many closed or exclusive groups that you will have a hard time joining (such as alumni groups from colleges and universities). LinkedIn had over ninety thousand groups a couple of years ago, some of which you might be able to join. When you join a group, you get access to the other group members. Group memberships can help you get in touch with people who share certain commonalities, such as geographic locations, associations, university affiliations, and interests. I have joined dozens of groups that give me immediate access to message any other group member. I've strategically joined groups with members I want to develop relationships with.

6. **Opportunity to show you are plugged in to current technology.** Having a LinkedIn account doesn't necessarily brand you as someone smart or technologically hip—but it can help! If you understand LinkedIn (and other tools) to some degree, you can communicate with others about these modern

tools and resources. Of course, LinkedIn won't make you smarter than you already are, but being able to talk about it intelligently and show that you have more than six connections will tell others that you are serious and competent about networking, new technology, and your career. In today's world, it's practically expected.

LinkedIn Limitations

As I mentioned earlier, the biggest problem with LinkedIn is not the technology, but the expectation people have that it will do great things for their networking efforts. By now, most professionals have heard of LinkedIn; they know it's highly recommended, and I think their expectations are high. Then, they get on the network and wait for the value to come. You have to be there, have an optimized profile, and even do some real work before you find significant value. To understand the kind of work you need to do, let's first understand what LinkedIn is NOT.

1. **LinkedIn is not a full-fledged social environment (although it seems to be playing catch-up).** LinkedIn has never been as feature-rich in the social space as other sites like Myspace and Facebook, which was just fine for many users. Most people wanted to find talent and relationships or to be found, not share pictures and videos from the weekend party. Myspace and Facebook got clogged up with a bunch of distracting junk while LinkedIn remained very clean, professional, and noise-free. During 2009–2010 we saw a lot of new social features, however, including major enhancements to the status update, group discussions, and companies. All of these changes bring us closer to more traditional social sharing, following, and commenting. The question is, how much is enough and how much is too much?

2. **LinkedIn does not represent your entire network.** Your entire network CANNOT be on LinkedIn, unless you never go out and never talk on the phone. Are your plumber and your mechanic connected to you on LinkedIn? Mine aren't. Is everyone from your family reunion on LinkedIn connected to you? I didn't think so. Don't get hung up on the idea that your network on LinkedIn, which might have fifty first-degree contacts or five thousand first-degree contacts, is your entire network. This is just a subset of your overall network.

3. **LinkedIn does not give you complete control of the relation-ships you have.** I've made the point that you can only connect to me if:

   a. you are on LinkedIn, and

   b. I agree to be your connection.

   What if you are not on LinkedIn? We'll never connect (unless I convince you to get on LinkedIn, but I'm not going around selling everyone on the virtues of getting a LinkedIn account). You simply cannot have every single person you know as a first-degree contact. To drive the point home, if I want to disconnect from you, I can do it with just a few clicks. You don't have any control over whether I stay in your network or not. That's why a complementary relationship management tool comes in handy, since you can put whomever you want into that database.

4. **LinkedIn does not allow you to control or change any infor-mation on your contacts.** As mentioned, people commonly con-fuse LinkedIn as a relationship management, or customer relationship management (CRM), tool. CRM refers to software a salesperson uses to keep track of clients, prospects, and other contacts. Some of the most common CRM tools include Sales-force.com, GoldMine®, and ACT! There are hundreds of CRM tools available. My software, JibberJobber.com, was de-signed as a CRM for professionals in transition, but has been adopted by a number of businesses as their solopreneur CRM.

5. **LinkedIn does not allow you to store much private relationship information about your contacts.** When you go to your contacts list, you can drill down on a contact and store various information, such as phone numbers, notes, and tags. This is a part of CRM functionality. What's missing is the ability to store time-stamped log entries (instead of one big blob of notes) and, more importantly, to make associations between contacts and companies. For example, if I meet you at a networking event and find you work for one of my target companies, I can't go into your LinkedIn profile and associate you with that target company (I can include it in the notes, but I want to go to a company page and see all of my contacts associated to the company). Additionally, in LinkedIn, I can't create an action item, such as "Call Jim next Thursday to see if he talked to the VP of finance for me." Whether Jim is going to talk to the VP of finance to plug me

as a prospective employee or vendor, it is an important action item to follow up on! LinkedIn doesn't allow you to log meetings or thoughts, create action items, or even rank the relationship you have with me—all of which are important things to track! Finally, as mentioned earlier, even if you do track most of this stuff on each contact record, what if that contact disconnects from you? I'm still unclear what happens to the data, but I don't trust LinkedIn with my very important notes.

6. **LinkedIn doesn't really provide much privacy.** Information from your Public Profile will be available for people to see, no matter what you declare as "public" information. Your network will be available for others to see, even if you turn off "Browse Connections" (see the Account & Settings chapter). Privacy issues on LinkedIn include sensitive information your company might not want others to know, as well as those who want to perform a job search without their employer knowing. Having said all that, here's something I learned from an information security professional: if you are looking for privacy, stay away from the Internet—period!

7. **LinkedIn has a closed communication system.** I don't like it when a service requires me to log into its website in order to get information or communicate with someone who is reaching out to me. I would like to reply to LinkedIn InMail messages from my email, but I find many times I need to reply from within the InMail system. This protocol is common for web services such as Facebook and Twitter. I dislike this method because I feel like it disrespects my time. At least InMail shows me the actual message in my email box so I don't need to login just to see what the message says, but I wish I could quickly and easily reply from my email client.

There are other limitations. For example, some are upset at the limit on the number of first-degree contacts (thirty thousand), although I doubt that will affect more than a handful of people. I should note, after listing these limitations, I am not alone in thinking that LinkedIn **should not necessarily** resolve all (or any) of these things! While it would be cool to have a super-system, a silver-bullet to solve all of my networking needs, I do not advocate LinkedIn as the total solution. Here's why:

First, LinkedIn still needs to polish and fine-tune its core functionality. As a software developer, I understand the danger of creating more features while the core is either not complete or not architected to work with so many peripheral features. You may have used software that introduces cool bells and whistles that aren't relevant to you, while you still have complaints about the core functionality. It's always appealing to develop a broader offering, but it's usually not good for the end-users unless your core is really strong.

Second, aside from spreading itself too thin technically, it can be an enormous distraction to design, support, and maintain various additions to the core function. Designers need to make sure they do not break anything that exists (or should exist) in the core. The support team will need to learn new tools, philosophies, and rules. Introducing any new feature outside of the core can add exponential complexity with regard to maintaining the software and hardware. It's not to say that LinkedIn can't tackle any new non-core features, but there is a reason why companies in every industry stay within their core competency and outsource the rest.

Third, there is a new trend in software development, especially web-based applications, where you develop the very best program to accomplish A, B, C and then find a complementary Web-based application to accomplish X, Y, Z. This way, multiple companies share the burdens of designing, staying relevant, maintaining, etc. LinkedIn should be the very best application on which to find and be found and to portray your professional image (to a degree). It should find other systems that make up for its shortcomings and partner with them. Right now these are called "mashups," and there are thousands of examples. Here are a few:

- Most websites that require a login mash up with CAPTCHA technology, or something similar. For example, If you leave a comment on a TypePad® blog you'll likely have to type in the letters/numbers from an image to verify that you are a human being (and not an Internet robot). LinkedIn introduced this in some instances during 2010.

- **Google Maps** integrations with sites to find sex offenders, bars, transit lines, or hundreds of other useful things. GoogleMapsMania.com is a website with information on the Google Maps mashups.

- JibberJobber.com has mashups with various systems, including Google Maps to see your networks on a map, **Skype** to call a contact with one click, **Anagram**™ to quickly capture email signatures and add them to your network, and **Indeed.com** to search multiple job boards at once.

In June 2007, LinkedIn announced they were going to follow Facebook's lead and allow third-party companies to create applications for LinkedIn users (similar to the mashup idea). In the last edition of this book, I wrote: "I'm guessing they are going to restrict access to a handful of strategic partners. I hope I'm proven wrong."

So far, I wasn't wrong. Since LinkedIn Applications was announced, there have been nineteen approved applications (as of this writing, compared with tens of thousands of applications on Facebook). Some of the applications are immensely useful, others are confusing, and others are useful in a corporate or team setting. They are definitely restricting their APIs to a handful of strategic partners. I think this is a good call because, ultimately, any application you get from LinkedIn has its stamp of approval, which means it has probably gone through a vigorous quality assurance process.

Perhaps even more exciting is the API access that has matured since they announced it. I've seen some exciting interfaces from third-party websites (outside of LinkedIn) and expect that to get much richer and more common during 2011!

Chapter Summary

- LinkedIn, while a powerful tool, is not a silver bullet.

- LinkedIn should be a component of your networking strategy, complemented with other tools and techniques.

- Different people use LinkedIn for different reasons—why do you use it?

- There are a lot of exciting changes with LinkedIn interfacing with third-party sites, and more are on the horizon!

"I'm a huge fan, user, and promoter of LinkedIn. I've used it to network into the technology field from print advertising and get my current job in a new high-growth department. Currently, a friend and I are founding an online apparel business. I've used LinkedIn to find manufacturers, screen printers, and as a resource for investors to research our management team."
Jimmy Hendricks, Cofounder and CEO, Deal Current,
http://www.dealcurrent.com

"So far, I primarily see my involvement on LinkedIn similar to having money in a savings account. It's not something I often tap, but something that's reassuring to have available when the need arises. I've also used LinkedIn as a journalist to make contact with people who might have expert knowledge on a topic."
Bernie Wagenblast, Editor, *Transportation Communications Newsletter,* http://transport-communications.blogspot.com/

"My favorite thing about LinkedIn is finding old contacts. I discovered an old college friend and several coworkers who moved on to other companies. It was nice to read about what they'd been up to in their profiles and reconnect."
Pete Johnson, Chief Architect, Hewlett-Packard,
http://www.hp.com/

"It's not about 'collecting and trading your friends' as I originally saw LinkedIn. It's really about visibility into relationships that you wouldn't know existed otherwise. The more people who know how to use the tool effectively, the more effective it is for everyone. What a terrific way to take your relationship to the next level!"
Scott Ingram, CEO and Founder, Network In Austin,
http://www.networkinaustin.com

---

3 Your Profile

This chapter talks about optimizing your profile to

1. increase your chances of being found, and
2. effectively communicate information about you.

I remember something I heard about resumes: your resume is not your obituary, it is a marketing tool to get you into an interview. Job seekers are inclined to put all of their history on a resume, worried that leaving something off might mean they don't get a job (or an interview).

But the purpose of the resume is not to share all of the great things you've ever done.

It's the same thing with your LinkedIn profile. This is a chance to market you, your products, and your services. If you clutter it up with factoids that are not important to your marketing message, you are only distracting the reader.

Your profile should communicate your professional brand, including your strengths, knowledge, and experience, if it is relevant to the role or services people are looking for (or, for the brand that you are building). Here are some things to consider about your profile:

1. **How will old friends and people who knew you from your past find you?** Make sure you put names of schools, companies, and clubs in your profile. You never know when that friend from five companies ago might search for you.

2. **How will recruiters, hiring managers, or potential business partners find you?** What will they look for? Usually they aren't searching for your name. They might search for the names of schools you attended, companies (even companies that went away through a merger/acquisition), and clubs or associations. Recruiters look for candidates with specific experience, affiliations, or work history. Include keywords and jargon you would include on a resume so a recruiter looking for a "PHP programmer with CSS and Adobe skills" or a "project manager with a PMP" can be found. Many recruiters have had training on how to search on LinkedIn and use various search techniques to find exactly what they are looking for, based on keywords, companies, etc.

3. **What first impression does your profile give?** You should spend at least as much time on your LinkedIn profile as you would on your resume. Whether you are looking for a job or not, you never know who is looking for you. Ensure you leave a sharp first impression by having proper grammar and spelling in your profile. Does your profile tell your story enough to engage me when I read it? Is it interesting, and do I walk away thinking, "This is the right person! My search is over!"

People may look for you based on things you have in common, like where you went to school. For example, let's say you went to college at UCLA. You might have a classmate who has been looking for you but can't remember your name. When he looks for other UCLA alumni, he might recognize your name or picture.

At the bottom of the homepage is a "Just joined LinkedIn" section which has my previous companies and universities (from my profile). Once, I clicked on a company I worked at many years earlier. One of my key contacts from that company had gone on to a different company, but because he had put the company name in his profile, I was able to find him, and we've since reconnected!

Unlike a resume, which is limited in the number of pages (usually two), a LinkedIn profile can be very, very long—and that is OKAY! Just make sure the length doesn't take away from the marketing message. This might seem at odds, but consider this: your marketing message should come across very clear, and even concise. How can you do that with a long profile? You would only work towards a long profile if you are trying to get more keywords, companies, etc. on your profile. Think about what your target contact would search for when looking for someone of your talent and make sure those keywords and phrases are peppered throughout your profile.

Many people think a LinkedIn profile is their new resume. I do not agree, as I don't think anything is going to replace the traditional resume any time soon. The beauty of a LinkedIn profile is that it can be a resume whether you are in a job search or not. In other words, even if you aren't in an active transition you can still let others know about you and what you have to offer. Many professionals who are not in transition have LinkedIn profiles.

A resume has various information, including your job history, education, and skills. LinkedIn has sections for this information and more. Here are some differences between a LinkedIn profile and a traditional resume:

1. Currently LinkedIn doesn't allow bulleted formatting. You can get around this with a trick, but it still isn't the regular bullet formatting that we're used to. Because a profile doesn't have bullets, people tend to write in a paragraph narrative format, without the typical "quantification" you see on a resume. But if you are good, you'll include proof showing how valuable you were to a particular company. For example, on your resume you might put statements, on each bullet point, like "increased department revenues by 150 percent and profits by 250 percent in 18 months." You can include that in a LinkedIn profile, but it seems most people don't.

2. LinkedIn has a section at the bottom of each profile where you can state your "interests," with predefined interests such as career opportunities, consulting offers, new ventures, job inquiries,

and a few others. This helps people who view your profile know whether you are approachable for certain things or not. Some people will respect your list of interests while others may still attempt to contact you for something, even though you do not list it. (Not everyone who is interested in career opportunities is going to put it on the list, especially if their current employer doesn't know they are open to other opportunities!)

3. If you are a member of any LinkedIn groups, the groups will show up on your profile (you can choose to not show any groups, but by default they will show). You can learn more about this in the "LinkedIn Groups" chapter.

One of the most important considerations while creating your profile is this: how will people, using the LinkedIn search box, find (or miss!) your profile? Also, the Public Profile is indexed by search engines such as Yahoo! and Google. The way you optimize your profile for the LinkedIn search function also applies to other search engines.

There is a common phrase amongst Internet marketers called "search engine optimization" (SEO). The idea behind SEO is that you are trying to optimize a web page (or profile), so search engines bring it up first (or, in the first page of results). Optimize your LinkedIn profile with SEO in mind. Here are some tricks you can incorporate as you develop your profile:

1. **School names**—Include the full name (University of Virginia) as well as the common abbreviation (UVA).

2. **Company names**—Just as you did with the school names, make sure you put the official and common names of the companies where you worked. If your company is a subsidiary of a larger company, put the name of the larger company also. That way if a recruiter is looking for someone from either the main company or the subsidiary, they are more likely to come across your profile.

3. **Technical skills**—Just as a recruiter looks for "project managers" or someone who is a "project manager professional," they might search for "PMP" or someone who is a member of the "PMI" (Project Management Institute).

4. **Keywords**—Include keywords used to describe your skills, abilities, or professional passions. I've found that putting the keyword as your current title increases your SEO, and I've heard increasing the number of occurrences of a keyword (or key phrase) increases your SEO.

Aside from writing your profile so the search engine finds you, consider writing it so the person reading your profile is impressed by the value you bring to a relationship. Deb Dib, The CEO Coach, wrote a powerful article for CEOs titled "LinkedIn—What It Is and Why You Need to Be On It."[7] Check out the eight profiles she links to, which are all examples of excellent LinkedIn profiles, whether you are a CEO or not!

TIP: Lonny Gulden, a recruiter in Minnesota, suggests including company names from any company you worked with but has since changed its name (merger, acquisition, etc.). Recruiters might look for ex-employees of those defunct companies.

Perhaps one of the best resources for optimizing your profile so it appeals to people *and* LinkedIn's search algorithm is Guy Kawasaki's famous "LinkedIn Profile Extreme Makeover" (you can find it by typing "Kawasaki & LinkedIn" into a search engine, or by going to http://tinyurl.com/KawasakiMakeover (blog.guykawasaki.com/2007/01/linkedin_profil.html).
Guy probably has one of the most viewed profiles on LinkedIn. The tips in the makeover come directly from LinkedIn.

LinkedIn advises us to flesh out each section of our profile and put in more details (which gives us a chance to include more keywords). Some of their recommendations include:

- "Write recommendations"—This is a way to get YOUR NAME, with a link back to your LinkedIn profile, on someone else's profile. It's like getting a little bumper sticker on their profile linking right back to your profile.

---

7. http://tinyurl.com/ExecLinkedIn
   job-hunt.org/executive-job-search/linkedin-for-executives.shtml

- "Ask a question, answer questions"—Again, this is a way to get a link to your profile on a totally different section of LinkedIn. It is also a chance to get your brand and knowledge in front of people outside of your network.

- "Get a Vanity URL"—This is quick and easy and something you should do right now (seriously, right now). Click on "Profile" from the main menu, then click on the "Edit" link by your Public Profile, and then click on "Edit" next to your Public Profile URL. You should modify the URL from something like
  http://linkedin.com/pub/1/234/1344
  to http://linkedin.com/in/jasonalba, which looks more "on purpose."

- In the Summary section, "LinkedIn said to add substance...this is your elevator pitch." I encourage my clients to fill in as much of their summary as they can. You get two thousand characters; I say use as many as you can!

And on and on. Make sure you read the original post to see what LinkedIn recommended. When I critique a profile, I typically look at many things, including:

1. **Name:** Is it just your name or does it include superfluous information like a phone number, email address, or credentials and acronyms? In general, with few exceptions, it should just be your name.

2. **Picture:** Use a close-up of a headshot that looks professional. The picture makes your profile more personable.

3. **Professional headline:** This is YOUR CHANCE to share your value proposition. Don't just put title/company, like so many people do. Tell me what value you bring to me (or, to your client or prospect).

4. **Current title:** As mentioned above, this is where you put key phrases that people might search for.

5. **Websites:** You should have three websites listed and the descriptive text for each one so even if I don't click on the links I still get your brand messaging.

6. **Vanity URL:** As mentioned above, edit your Public Profile URL so it has your name, or something more deliberate.

7. **Summary:** Use as many of the two thousand characters as possible, and write something engaging! You can use Problem, Action, Results (PAR) statements, or something clever. Spend time on the Summary and get it right!

8. **Experience and education:** This is another opportunity to include keywords for SEO as well as stories and supportive brand messaging.

9. **Interests:** Below Recommendations is the Interests section—yet another area to freehand more information, branding, and SEO words.

10. **Groups and associations:** ONLY show groups that are supportive of your brand. You don't need to show all groups you are a member of—only show those that support your brand.

11. **Contact settings:** This is an excellent place to put something like, "The best way to contact me is via email at <u>Jason@JibberJobber.com</u>," or something like that. This helps people contact you even if they aren't connected to you.

Once you get a strong profile, you don't have to revisit it to keep it up to date. Of course, you should keep it updated with relevant changes, but most people can easily go a year or two (or more) without updating their profile. Scott Allen, a social marketing expert, shared a story on his blog[8] about a timely profile update that resulted in $5,000 of additional revenue.

Your Public Profile is what people see if they are not logged into LinkedIn. It's important to think about what is visible and what is hidden. Click on "Settings" and then "Public Profile" on the Settings page to choose what you show and hide, managed from this screen:

---

8. <u>http://tinyurl.com/5000Profile</u>
linkedintelligence.com/the-5000-profile-update/

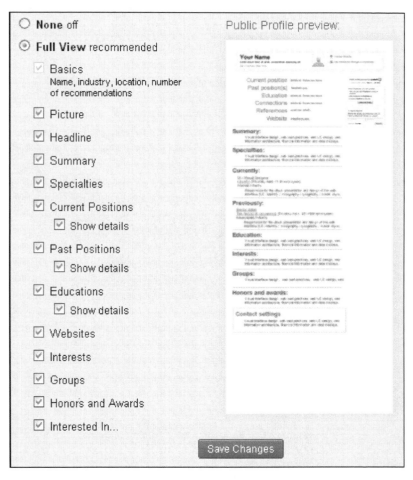

○ **None** off

◉ **Full View** recommended

☑ Basics
Name, industry, location, number
of recommendations

☑ Picture

☑ Headline

☑ Summary

☑ Specialties

☑ Current Positions
　　☑ Show details

☑ Past Positions
　　☑ Show details

☑ Educations
　　☑ Show details

☑ Websites

☑ Interests

☑ Groups

☑ Honors and Awards

☑ Interested In...

Public Profile preview:

Save Changes

Figure 1: Public Profile Preview

This screen allows you to choose what information people can see if they are not logged in to LinkedIn. In other words, it allows you to choose the degree of privacy. I have two thoughts on this:

First, you should provide as much information as possible so visitors who find your profile can get the gist of who you are, and what you have to offer, without having to login (or worse, create a new account and then login).

Second, consider your personal comfort level regarding privacy. If you are comfortable putting up certain information on your regular, non-public profile, which is available to over eighty million people, why not let non-LinkedIn people see it? It doesn't make sense to hide anything from people who are not logged into LinkedIn, especially considering you shouldn't put stuff up that is not on-brand.

I regularly hear concerns about privacy with a LinkedIn profile. The purpose of a profile is not to record private data—it is to showcase who you are in a professional setting. It's great to be concerned about privacy, but with the technology available today, there are dozens of places someone can go to find your personal information, including your contact information, professional/work history, home address and telephone number, even your social security number! My recommendation is to click every box in the image below and let everyone see what you put on your LinkedIn profile.

Having a strong and well-crafted profile gives you the opportunity to communicate your brand, abilities, and professional goals. The alternative is to let search engines determine that for you (you want more control over your brand and messaging, don't you?) by listing whatever pages they have for your name. Your profile, and the ability to be found by people who are looking for you, is central to your success with LinkedIn, so make sure you spend enough time and effort to get it right.

Chapter Summary

- Set up your profile as completely as possible so that others have a better chance to find you. Don't worry about the profile completeness percentage on your profile page, as this has a tendency to change when LinkedIn introduces new features.

- Make sure you check spelling, grammar, and overall readability of your profile—it's not just for the search engine but for the human who reads your profile.

- Take advantage of things such as the Vanity URL and recommendations to make your profile look stronger and more professional.

- Allow others to see all of the information on your Public Profile.

"I'm starting to think LinkedIn may be Resume 2.0, but I'm not 100 percent convinced yet. When I beefed up my profile, I actually copied and pasted bullet points from my latest conventional resume, and now that thing is searchable for all the world to see. That's a lot different than posting it in some random place on a few job sites. It's in a neutral place where you expect pressure-reduced interactions with others."

Pete Johnson, Chief Architect, Hewlett-Packard,
http://www.hp.com

"You only get three links, so use them wisely—I would encourage you to use the links for your blog, a full page of other endorsements about you, and your portfolio or book or something else that shows your expertise."

Phil Gerbyshak, Author, *10 Ways to Make It Great!*; Owner, The Make It Great Guy, http://www.makeitgreatguy.com

"When you set up your profile, make sure it's a snapshot—no one wants to read a detailed autobiography. And don't forget to proofread!"

Christine Dennison, Job Search Expert, Dennison Career Services, http://www.thejobsearchcoach.com

"Your LinkedIn profile is a critical part of your online identity. Make sure to complete it with information that conveys what you do, the value you deliver, and the audience you serve; in other words, have it reflect your personal brand."

Walter Akana, Life Strategist, Threshold Consulting,
http://www.threshold-consulting.com

"I think the status update feature is one of the coolest things on the LinkedIn profile! Used strategically and well, the status update line can be used to promote your business, products, and personal brand to your network. I use it to promote my blog posts, speaking engagements, and networking events or conferences I'm attending to facilitate the possibility of connecting in person or on the Web with those in my network. As a recruiter, I also use the status update to make my network aware of opportunities that I'm recruiting for. I have a fairly large network—over 1,500 connections—so that's 1,500+ opportunities to reach candidates or get referrals—and it works! I also encourage job seekers to use the status update feature to make sure their network knows that they are currently searching for a new opportunity. Regularly updating your status keeps you on 'top of mind' and on the homepage of those in your network."
Jennifer McClure, Executive Recruiter/Executive Coach; and President, Unbridled Talent, LLC, http://unbridledtalent.com

4 Account & Settings

There is a "Settings" link under the dropdown at the top right of every page (just click your name to see the Settings link).

Figure 2: Account Settings

The Account & Settings page takes you to one of the most important menus in LinkedIn since this is where you can determine what messages you get via email. The Account section shows you what you get at your current account level and allows you to see what you could get if you upgrade. On my page, it shows I have four of five Introductions left and no InMails.

Currently there are four different types of accounts: Basic (free), Business, Business Plus, and Executive. You can see a comparison chart showing features such as the number of InMails you can send as well as what and how many profiles you can see on searches. Refer to the chart to see the most current benefits of upgrading.

Figure 3: Account Types

Unless you have a very specific reason to use LinkedIn as a prospecting database, which means you need the best possible results on your searches, and quick access to message people you aren't connected to, a free Basic Account should be just fine. Who would use LinkedIn as a prospecting database? Recruiter and sales professionals and maybe job seekers who are looking for higher-level positions are examples of people who would find value in upgrading. If you are not spending much time on LinkedIn, not searching a lot, or not trying to communicate with people you don't find in search results, you should be fine with a free account.

Settings is one of the most important places to visit if you feel like you get too much or too little correspondence (emails) from LinkedIn. There are different functions in LinkedIn that can send communications to you. For example, if someone wants to get in touch with you, you get an email from LinkedIn. If someone asks a question in Answers, and you are in their network, LinkedIn might send you an email (see the chapter on Answers for more on this powerful feature).

The default settings did not meet my needs and might not meet yours. I was getting more emails than I wanted from LinkedIn (which didn't help my efforts to manage my increasing "information overload" I was getting from email, Facebook, Yahoo! Groups and other sources).

Frustrated by too many notifications, I finally went to the Settings section and changed the default settings so now I only get what I want in my email inbox (the rest are still available, if I want to see them, in my LinkedIn Inbox). I'm not limiting my ability to get messages; I'm simply choosing whether I get them via email or in my LinkedIn Inbox.

Email Notifications

Contact Settings
  You are receiving **Introductions and InMails.**

Receiving Messages
  Control how you receive emails and notifications.

Figure 4: Message Settings

Here's how I have my settings:

- **Introductions and InMails**—I get these via email, so if someone tries to get in touch with me I won't miss their message. The LinkedIn landing page is too cluttered for me to read every single message I get, and I don't want these types of messages to get lost. I also don't use LinkedIn's Inbox to manage any of my messages and don't want to have to go there to find these messages.

- **Invitations**—I also want invitations sent to my email inbox. An invitation comes from someone who wants to join my network and could be an important new relationship (or an important offline relationship that is ready to connect online).

- **Profile Updates**—At the time of this writing, this is no longer available via email. Perhaps it was being abused, or LinkedIn got too many complaints. Nonetheless, you will only be able to see updates on your first-degree contacts if you go into LinkedIn, and you won't be able to push notifications out just by changing or updating your profile.

- **Job Notifications**—These are job openings that come from people in my network. I thought it would be useful to get these notifications and pass leads to friends who are in transition. Unfortunately, I got so many irrelevant job notifications sent to me it was overwhelming. This was the main communication I got that prompted me to check out this place, and I quickly changed it from "Individual Email" to "No Email."

- **Questions from your Connections**—My network has grown to a size that produces some negative side effects, one of which is that I would get way too many emails of new questions from LinkedIn Answers from my first-degree contacts. I simply go to the LinkedIn Answers page and look for relevant questions I can respond to from there, rather than have my Inbox clogged up.

Since the last edition of this book, there are some email/web settings that are new and others that have dropped off. If you find yourself getting too many emails from LinkedIn, simply go to the Receiving Messages section to see if there is a new setting you need to change.

Also under Settings is the "Connections Browse" option, under **Privacy Settings**. This has become a controversial option. You can choose to show your connection list or to hide your connection list from your first-degree contacts.

Hiding your connections list seems to be contradictory to what LinkedIn is about: the ability to see who is in my contact's network! However, the option is a privacy setting and I've heard from people who have needed that level of "supposed privacy." Why do I say "supposed privacy?" Even if you choose to hide your connections list, your first-degree contacts can still see who you are connected to!

The default setting shows your connections list, which is a page formatted to show all of your first-degree contacts in an easy-to-digest view. If you change the default to hide the connection list, your contacts can still see other first-degree contacts through a simple search.

For example, let's say you have a first-degree contact who is a project manager in Seattle. When I search for "project manager" or "Seattle" I will see everyone in my first three degrees (including your first- and second-degree contacts), **even if** you have said I can't browse your network. Notice the difference between "browse" and "see." Browsing shows me a nicely formatted page with your first-degree contacts. Seeing is just the ability to find read a profile. You cannot block me from seeing people in your network.

Another critical setting has to do with your email address(es). I encourage you to ensure you have the right email addresses associated with your account. I have two email addresses I use: one for the business I own and another through Gmail. I send a lot of emails each day, some from my business account and others from my Gmail account. Because of this, I make sure people who want to invite me can use either address. Why? Some of my contacts only know me from the Gmail account while others know me from my business account. Fortunately, LinkedIn allows potential contacts to send an invitation to either address.

Figure 5: Email Accounts

It is essential you use an email address you own, and have complete control over, as the primary email address. I saw an email thread where the discussion was based around a person developing a LinkedIn network while working at their job and the employer claiming the network belonged to the company. This is an interesting claim. I can see the employer claiming the LinkedIn account and network (and network connections).

To avoid this potential conflict, make sure you control the email addresses on your account. You can get free email accounts from Hotmail, Yahoo!, Gmail and a host of other services. Also, make sure your primary address is one of YOUR email addresses, not a company address, even if you own the company.

Here's a summary of my settings:

Setting Name	How I've Set My Account
Introductions and InMails	Email me immediately
Invitations	Email me immediately
Job Notifications	No email; I'll look on the website
Answers Notifications—new questions	No email; I'll look on the website
Connections Browse	Allow others to browse my connections
Email Addresses	My Gmail account is the primary address, my JibberJobber account is a secondary address

Chapter Summary

- Dive into the Account & Settings pages to set up your preferences—this will determine when you receive emails and when you don't!

- If you have multiple email addresses make sure you put them in your account—that way others can connect with you by sending an invitation to either email account. Never use an employer-sponsored email as your primary email on record.

- Turning off the Network Browse can prevent your first-degree contacts from seeing the complete list of all of your first-degree contacts, but does not block those contacts from showing up in search results.

"Learn the site. There is a lot of useful help listed under Help & FAQ. If you take the time to review your Account & Settings and set your account up properly, your time on LinkedIn will be much easier."
Sheilah Etheridge, Owner, SME Management,
http://www.linkedin.com/in/smemanagement

5 Connecting with Others

Central to your LinkedIn experience, and your success on LinkedIn, is how and when to connect with other people. Consider a spectrum where one end represents people who are open networkers—they accept any invitation they get and freely extend invitations to anyone they can. On the other end of the spectrum are people who are more conservative about whom they connect with—the closed networker.

I can't tell you what kind of networker you should be—it's different for everyone. Depending on your circumstances, you might choose to change your connection strategy over time. In this chapter, we'll talk about considerations regarding the open/closed spectrum concept. I'll explain characteristics of both ends of the spectrum and give you enough information to decide what your own connection policy should be.

Open Networking

At the extreme of open networking are those who connect with anyone, no matter what. You may have heard of the LION, which is an acronym that stands for "LinkedIn Open Networker." I wrote a blog post titled "I'm a LION—Hear Me ROAR"[9] that got a number of interesting comments. LinkedIn LION refers to a person interested in having as many connections as they can get and indicates a supposed willingness to accept an invitation from anyone.

Not every LION is really an open networker, and not every open networker is a LION. Open networkers tend to subscribe to the theory that having more connections means you have more channels to reach a key person. This might be true (I'll share my personal connection philosophy at the end of this chapter), but I'm not convinced spending a lot of time haphazardly growing your network is a good use of time.

Why would you want to be an open networker? Perhaps having a large number of first-degree contacts gives me richer search results and allows me to communicate with more people.

For example, a recruiter wants to have access to "passive candidates" (people who are not active job seekers). If he has one hundred first-degree contacts in LinkedIn, he might find the right candidate for a job he's trying to fill. Imagine he has four thousand first-degree connections. Don't you think it's more likely he finds the right candidates when he searches through his network? A network search returns first-, second-, and third-degree contacts. This means every first-degree contact brings their network to you. This is why many recruiters on LinkedIn are "open networkers."

I used a recruiter for that example but there are other professions that find open networking beneficial. If you are in a roles where you actively look for people, or try to reach out to people, you would likely benefit from an open networking strategy. Some professions who would benefit from such a strategy include salespeople, business development people, entrepreneurs, or other power connectors (a phrase I was introduced to when I read Keith Ferrazzi's *Never Eat Alone*. He uses it to describe people who are in a profession where networking is a natural part of the job).

---

9. http://tinyurl.com/linkedin-LION
   imonlinkedinnowwhat.com/2008/07/31/im-a-lion-hear-me-roar/

Open networkers can quickly and easily grow their personal LinkedIn networks by downloading the Outlook toolbar and inviting all of their Outlook connections to join LinkedIn. In addition you can import your contact lists from online email accounts, such as Yahoo!, AOL, Hotmail, and Gmail.

I DO NOT RECOMMEND EITHER OF THOSE OPTIONS, even if you are a LION.

I recommend you prepare your contacts to connect with you, outside of LinkedIn, before you invite them through LinkedIn. In other words, ask them if they'd accept an invitation to connect in an email, over the phone, face-to-face, or whatever before you send an invitation to connect.

Historically, each user has had a limit of invitations they can send out (the limit has been three thousand). I've never reached this limit, but I'm regularly asked about it. If you use your three thousand invitations, you have, in the past, been able to request more from LinkedIn customer service. Additional invitations have been granted in blocks of one hundred.

There has been justified confusion amongst LinkedIn users about whether LinkedIn encourages you to be an open networker (based on the tools they provide and the ease of inviting a lot of people) or to be a closed networker. In the past, someone you invited could have clicked a button that said, "I don't know" you. If five people clicked that your account would have been penalized. This button is not there anymore, but many people are still gun-shy about whom they invite.

Closed Networking

On the other end of the open/closed networker spectrum are the closed networkers. Closed networkers are those who only connect with those they "know and trust," which is what LinkedIn recommends (http://tinyurl.com/ClosedNetwork[10]). Instead of accepting every

---

10. http://tinyurl.com/ClosedNetwork
   help.linkedin.com/app/answers/detail/a_id/1303/kw/know%20trust

single connection request, they will evaluate the relationship with the person requesting the connection and only connect when they feel comfortable connecting.

Is it bad to be a closed networker? Not necessarily. There is value in knowing the people you are connected with. You can give more of your connections a recommendation than an open networker can. You are more likely to get real recommendations. You can feel comfortable passing introductions on because you know all of the parties in the Introduction (the person asking for the introduction and the person to be introduced to). Closed networkers value the individual relationship with each connection they have and treat LinkedIn as less of a database than open networkers do.

Many people who are closed networkers on LinkedIn are open networkers in a more general sense. Many will have more relaxed standards in other social networking sites or face-to-face networking. Somehow they think they really only should connect with people they know and trust. Many closed networkers use LinkedIn the way it was designed to be used, but might not get as much value out of their extended network (or the database) as an open networker could.

The Canned Invitation

Another LinkedIn topic that gets a lot of attention is how you ask someone to connect with you. Is there a cheesier way to invite someone to LinkedIn than using the default invitation that LinkedIn provides? When you invite someone to LinkedIn there is a default invitation you can change, but you have to remember to change it—and it's easy to miss! For many years the default invitation has been:

I'd like to add you to my professional network on LinkedIn.

Every blog post I've read regarding the default invitation is negative. Every LinkedIn consultant recommends you make the invitation personal. I totally agree! Leaving the invitation text as the canned text is like saying, "I don't care about you—you are just a number to me."

When I invite someone to connect in LinkedIn, I change the default invitation so the recipient knows (or remembers) who I am, and they get the feeling that I am serious about a professional relationship.

Incorporate your style in the invitations you send through LinkedIn—whether that is more casual or more business, your personality has a place in your communications.

Scott Allen, the social media expert, has a tongue-in-cheek blog post[11] sharing various types of creative invitations that are quite fun. The best invitations I receive are concise and include the following information (in any order):

- **Here's who I am.** This is not a time to brag or list all of your wonderful skills and experiences. Instead, let me know something about you without making me go look at your profile. Give our potential relationship context.

- **Here's how I came across your profile.** Did you find me in a LinkedIn search? Did you meet me at a networking event or find me on a Yahoo! Group? Let me know how you found me so I don't have to wonder if I'm just a number to you.

- **Here's why I'd like to connect.** You don't HAVE TO have a reason to connect, other than just wanting to connect and get to know one another better. However, if you tell me why you want to connect, I'll be able to determine what next steps might be. Every relationship starts somewhere. If you have a specific reason to connect, let me know what it is, and let me know how we can move towards that.

Earlier I asked if there was a cheesier way to invite someone to LinkedIn rather than using the default invitation that LinkedIn provides. Indeed, there is, and I see it regularly. Some people make equally ambiguous invitation messages and use them as their canned copy-and-paste messages, time and again. They don't customize it to include anything specific for OUR relationship. They replace one canned invitation with another canned invitation.

---

11. http://tinyurl.com/LinkedInvites
linkedintelligence.com/better-boilerplates/

YOUR Connection Strategy

The open/closed spectrum is just that: a spectrum. You might not be on either extreme, as defined above. Most people tend to sit somewhere in the middle. I won't say you are right or wrong—that is your decision.

In *The Virtual Handshake*, the authors talk about ensuring you have diversity in your network. Think about what that means—if you are an accountant, connected only with other accountants, you don't have diversity. What kind of professionals do accountants work with? CFOs, CEOs, accounting vendors, tax professionals, HR professionals, IT professionals, and more. See the diversity that accountants could have in their networks? What types of professionals should you include in your network?

In addition to adding diversity to your network with other types of professions, think about adding diversity with other industries. What contacts could you connect with from complementary industries? Think about your industry and who serves you—who are your vendors and service providers? Think about whom they work with—perhaps clients in other industries.

Also, who are your clients? Perhaps they could be categorized in other industries. As you include these people in your LinkedIn network, your diversity will grow in your first-degree network, and you should see a remarkable difference in your second and third degrees.

Finally, think about where your connections are. Do you do business locally? You should have a significant amount of local contacts. Do you do a lot of business in Seattle or Florida? Grow your networks where your clients and prospects are, even if the new contacts are not in your industry or profession.

Final Thoughts Regarding Connections

Issues surrounding invitations and connections are well debated. Some people are offended by those who don't share their views on how to use LinkedIn to network, and others pass judgment on those who have a different style (even though their objectives may be different).

Awareness of how others use LinkedIn and where they lie on the open/closed spectrum will help you understand why they may accept or reject a connection invitation, or even how they react to your request for an introduction. One of the most frequent questions I get is, "If I disconnect from someone, will they get a notice?" No one gets a notice when you disconnect from them—so feel free to clean up your network as often as you wish.

Chapter Summary

- There are pros and cons to open and closed networking strategies —you need to determine what's best for you.

- When you invite someone to connect, be sensitive to the invitation message and customize it when appropriate.

- Change your connection strategy as needed, and respect the strategies that others have chosen.

"LinkedIn works best when you use it with focused outcomes in mind. I am always open to invitations, but take a look at my profile first and tell me why you are interested in connecting with me. A little 'wooing' goes a long way!"
Garland Coulson, The EBusiness Tutor; and Founder, Free Traffic Bar, http://www.freetrafficbar.com

"I asked people to join my network due to their 'status'...what a mistake. Invite only people you trust, regardless of their position or status."
David Armstrong, Founder, Bounce Base,
http://davidarmstrong.me

"Most of my LinkedIn contacts come through an electronic newsletter I edit that's focused on the transportation industry. I sent an invitation to the subscribers who were already part of LinkedIn to connect with me. I see it as one more way readers of the newsletter might gain value from being subscribers. Since the subscriber list is private, those who are part of my LinkedIn network can see others who are part of my network and make a connection, either directly or through me."
Bernie Wagenblast, Editor, *Transportation Communications Newsletter*, http://transport-communications.blogspot.com/

"I wasted several InMails and Introductions when I could have simply hit the 'Add [contact name] to your Network' button because I thought I had to know their current emails to use that. I think this is very common."
Ingo Dean, Senior Manager of Global IT, Virage Logic,
http://www.linkedin.com/in/ingodean

"Be open to a diverse network. You never know whom you can help or who can help you. It's OK if you only want known people in your network, or only people from one industry, etc. But the more diverse your network is, the more rewarding it can be. This is especially true if you want to gain business from the site. People in your industry most likely won't need your services, but those in other industries may."
Sheilah Etheridge, Owner, SME Management,
http://www.linkedin.com/in/smemanagement

"Links to people can be broken—if you connected to a person that is suddenly sending you a boatload of spam just because you're 'LinkedIn buddies,' you can (and should) break the link."
Phil Gerbyshak, Author, *10 Ways to Make It Great!;* Owner, The Make It Great Guy, http://www.makeitgreatguy.com

"I hate receiving a LinkedIn 'invitation to connect' template. They lack authenticity and make me feel the person does not value the relationship enough to compose a personalized message. Take the time to create a touch point with your contact...it is well worth the extra effort."
Barbara Safani, Owner, Career Solvers,
http://www.careersolvers.com

Part II
Making It Work for You

Now that you have your profile and account settings set up, let's get into the proactive part of LinkedIn. In Part II, we'll talk about what it means to connect with other LinkedIn users, how to find relevant contacts, what "degrees of separation" is, how to give and accept recommendations, how to use the Jobs and Services section (even if you aren't in a job search), how LinkedIn Groups could benefit you, and what you should be doing with LinkedIn Answers.

6 Finding Contacts

Searching for contacts on LinkedIn was my first great frustration. I had fewer than five people in my network when I searched for management jobs in Utah. I was totally surprised to see no results on this simple and general search! This chapter talks about having a better search experience than I had, as well as how to find relevant contacts outside of the search function.

Here's what I've learned since those early days when I had only five connections:

Increase the size of your network. If you are interested in people and opportunities in your city, look for more local connections. If you are interested in people and opportunities in your industry or profession, expand your LinkedIn network with industry and professional contacts. As your network grows with relevant connections, you are more likely to get relevant search results. Your search results are based on your connections.

Connect with a few super-connectors. Super-connectors are LinkedIn users who have a lot of connections—hundreds or thousands of connections. Connecting with super-connectors significantly increases your visibility in the system. They enhance your ability to find and to be found. You don't have to connect with very many super-connectors. Some of them are general connectors while others might be highly relevant to you (based on location, profession, or industry). Connecting with even two or three super-connectors could give you a significant boost in visibility. As you connect with super-connectors, try to get to know them. Learn how you can add value to them, and become more than just another connection.

Use the basic search forms. It is easy to overlook this function in LinkedIn, but it's a significant part of your experience. The search form allows you to search people, jobs, answers, your inbox, and groups, using whatever keywords you want. You should list keywords, perhaps words that have to do with your profession or industry, and see what comes up. You'll likely find some new potential contacts to reach out to:

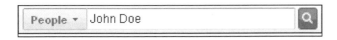

Figure 6: LinkedIn Basic Search

Understand advanced search options. Usually the quick search allows you to find what you're looking for. However, there is significant power in the advanced search page. For example, Scott Allen has talked about using LinkedIn to fill your spare time on a business trip by meeting new contacts you find on LinkedIn. Scott suggests you search for people in the city you will be in, like this:

Figure 7: LinkedIn Location Search

You can choose the country and zip code filter if you are in certain countries (including the United States, Canada, and United Kingdom). If you are an international traveler, you won't always be able to do this, but you might be able to get around the zip code search by using the country name in the keyword area.

Here are some of the useful features in the advanced search form:

Keywords—This could be a company name, a technology, the name of a certification, a club or school, or a first or last name. Hint: If you want to get results based on a company name, you should put that name in the Company field.

Title—I love this box. You know whom you want to network with, right—maybe not their name, but their title? When I was in my job search I wanted to network with "CEOs," "managers" and "project managers." Now that I'm in a sales role with my company, I'm looking for "career center directors." I leave all other fields blank and then fill in the Title box, and choose to limit the search to current titles only (in the drop-down menu).

Company—Do company research as you prepare to network into a company or research competitors. Try various versions of a company name (e.g., GE *and* General Electric) as well as names of subsidiaries, competitors, etc. You can limit the results to current companies only, or search for current and past companies, the same way you can with titles.

Location—Choose the country (and zip code for certain countries). I would complement a local search with other search criteria, such as "project manager" in Seattle (you'll need the zip code). This is a powerful way to find contacts that are local to you, or those who are in the same city you're traveling to.

Industry—Another favorite search of mine, leave all fields blank and select the industry you are interested in networking into. I don't know where the Industry list came from, but it doesn't seem to have changed in the last five years (some industries on the list are Research, Wine and Spirits, Machinery and Alternative Dispute Resolution). Results come from self-selection, which means people categorize themselves in an industry from their profile (on the page where you edit your name, professional headline, location, etc.). Because this list is not all-inclusive, I recommend you *put industry information in your summary*, which can increase your chances of being found.

School—This is a new field since the last edition of this book. I did a search on one of my alma maters and got different results when I used the school acronym (BYU: 1,454 results) than I did when I used the full name (Brigham Young University: 81,690 results). There are two lessons here. First, when you are searching for people by school, try it both ways. Second, make sure you have both versions in your profile.

Interested In—This is now a premium feature, unfortunately. This is pretty self-explanatory, but I don't put much value in it because I think people don't keep this section updated well.

NOTE: Any search criteria with a LinkedIn icon is not available unless you upgrade.

Sort By—This drop down menu allows you to order the results, with options such as "Relationship," "Relationship + Recommendations" and the number of "Connections." This helps you find something relevant to you without sifting through hundreds of irrelevant results.

The bigger your network, and the better connected they are (do they each have five connections, or do they each have fifty connections?), the richer results you should get.

Another LinkedIn search tip is to try Boolean searches (using AND, OR, and NOT), use quotes, inclusions, and exclusions. For example, try any of the following searches in the Advanced Search page (you can put these in the Keywords or Company or Title fields):

- Google NOT Microsoft (I got 37,661 results)

- Microsoft NOT Google (I got 164,131 results)

- Google AND Microsoft (I got 816 results)

- php -programmer -linux -javascript -css +html -mysql (a "+" means it has to include the phrase, "-" means the profile cannot include the phrase)

- "project management institute" (use quotes for an exact match of the term)

- "project management institute" AND (Microsoft OR SUN); (combine term in quotes, parenthesis, and AND/OR operators for very specific results)

This will be enough for most people on LinkedIn. For more in-depth information on really advanced searching, steal some tricks from Google's search tips page,[12] or check out http://www.booleanblackbelt.com.

In addition to using the search function, you should be able to find relevant contacts by your participation in LinkedIn Groups Discussions, as well as browsing through the group members. Look for people with whom you share common interests and reach out to them. You can also find relevant contacts in LinkedIn Answers, by asking questions relevant to them or by answer questions that others have posted. Finally, don't overlook the Jobs section as a place to find decision makers at various companies.

Chapter Summary

- Finding people is a major reason most people are on LinkedIn. Don't expect everyone to find you—take the initiative to look for them through various channels.

- The profiles you see in search results has been tied to the size of your network.

- The advanced search form allows you to really narrow down your search, based on industry, job title, company, location, and more.

- Use Boolean and advanced search techniques to narrow down the search results.

- Look for new contacts in other areas, such as Answers, Groups, and Jobs.

---

12. http://tinyurl.com/SearchBasics
google.com/support/websearch/bin/answer.py?
hl=en&answer=134479&rd=1

"Take the time to look through the networks of your direct connects. This is where you can easily find people you'd like to connect with, and you'll know you can ask your contact to help with the connection."
Scott Ingram, CEO and Founder, Network In Austin,
http://www.networkinaustin.com

"Be sure to search on the actual leadership competencies that matter instead of keywords like job title and company name. Some of the best and most interesting thought leaders of the future I've met on LinkedIn are the ones who haven't worked at cookie-cutter company X doing cookie-cutter job Y. The team you assemble won't be filled with the limiting beliefs of your competitors from several years ago, which is a major plus."
David Dalka, Senior Marketing and Business Development Professional, http://www.daviddalka.com

7 Understanding Degrees of Separation

The basic idea behind "degrees of separation" is that you are only so many degrees away from anyone (usually six). I've read that with technology (LinkedIn, Facebook, and Twitter) we are only three or four degrees away from anyone we want to network with.

For example, if I wanted to meet the prime minister of some obscure country, I should be able to network into him or her and only have to go to my fifth-degree connection before I get an introduction. That is, I talk to Mike (first degree) who says I should talk to Meagan (second degree) who says her friend Kim (third degree) knows someone; so she introduces me to Chuck (fourth degree) who introduces me to John (fifth degree) who happens to have a personal relationship with the prime minister (who is a sixth-degree contact).

A few years ago there was a "Six Degrees of Kevin Bacon"[13] experiment, and network television has done similar experiments (without the celebrity). It seems crazy, but you are probably

---

13. http://tinyurl.com/VI-Degrees
en.wikipedia.org/wiki/Six_Degrees_of_
Kevin_Bacon

just a few introductions away from talking to the person you've been trying to network into! The degrees-of-separation measurement that LinkedIn uses gives you an interesting perspective on your network (although it can be misleading).

When you look at the number of your first-degree contacts in your network, you can get a sense of the **breadth** (how wide) and **depth** (how deep) of your network. In other words, the width of your network is depicted by how many first-degree contacts you have, and the depth of your network is illustrated by anything below the first-degree.

On your landing page, you can see how many people are in the first three degrees of your network, like this:

1	**Your Connections** Your trusted friends and colleagues	4,009
2	**Two degrees away** Friends of friends; each connected to one of your connections	1,490,900+
3	**Three degrees away** Reach these users through a friend and one of their friends	14,305,100+
	Total users you can contact through an Introduction	15,800,000+

Figure 8: First Three Degrees of Connections

Figure 8 shows my network when I had 4,009 first-degree connections. These are people who have asked to connect with me or agreed to an invitation I've sent them. We have a mutual relationship because we are both first-degree contacts with one another.

Through those 4,009 contacts, I have a reach to 1.4 million people, who are my second-degree contacts. On average, each of my first-degree contacts has about 360 first-degree contacts, which are all second-degree contacts to me. I have a reach of fourteen million third-degree contacts (each of my second-degree contacts has, on average, ten first-degree contacts).

You have certain privileges with your first-degree contacts that you don't have with other contacts. For example, I can see a third-degree contact's profile, but I cannot see their email address. If I want to contact them, I have to use an Introduction request (or some other LinkedIn communication tool). Let's say I want to contact John Smith, who happens to be a third-degree contact.

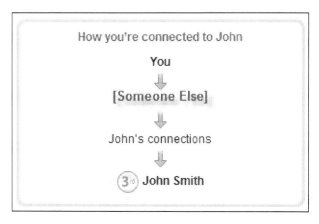

Figure 9: Third-Degree Contact

You can see I'm connected to "John Smith" through someone else (I took the name out for privacy). Instead of showing me John Smith's contact information (which I could see if we were first-degree contacts), it shows me who I know who knows him. I would contact the person I'm connected with through LinkedIn with a note that explains why I want to contact John, and an introduction note for John. My contact could then pass this on to his contact (my second-degree contact), and I am hopeful that my request is sent to John.

If each person in this chain is comfortable with our relationship, they will likely pass the requests on (this is where that "know and trust" comes in handy!). My note to John would carry some weight since it is coming through a trusted contact. Conceptually, this is a great idea, but there are two issues:

First, there is no guarantee I have strong relationships with my connections. I personally haven't met many of my first-degree contacts and might have some who would not send on an Introduction request.

Second, since the request travels through email, someone in the chain might not receive it in a timely manner. If it takes two days for each person to receive the email and forward it on, it might be six days before an Introduction request is even received by John! If you are in a hurry, it might be best to just pick up the phone (old-fashioned yet effective).

I encourage you to look at your first-degree contacts' connections to see if there are people you should have a first-degree relationship with. You could invite them to become a first-degree connection to you. I mentioned that the degrees of separation could me misleading, and here's why—if I ask John to join my network, we lose the history of how we knew one another (or, who was between him and me).

The degrees-of-separation measurement is helpful, but it's just one metric (which is somewhat faulty). I've seen people confuse the size, width, or depth of their network with the strength of their network. In addition to the size, they should think about (and measure) the strength of the relationship with each person in their network (which is not measured in LinkedIn right now).

Chapter Summary

- The "degrees of separation" concept gives you the ability to see how big your first-, second-, and third-degree networks are.

- When searching for new contacts, you can easily see the path between you and them.

8 Recommendations

LinkedIn Recommendations are third-party professional endorsements. Recommendations carry weight because you cannot give one to yourself. When you get a recommendation, you can choose to put it on your profile or not, but you can't edit it. Only the person who wrote the recommendation can edit it. This gives credibility to the recommendation.

Here's how recommendations work: find a contact you can recommend and go to their profile page. Click on the link that says, "Recommend this person." You have to say whether you are a colleague, service provider, business partner (which means neither of the other two), or had a student relationship (you were at the same school at the same time). I usually choose business partner since this is the closest description for most of my contacts.

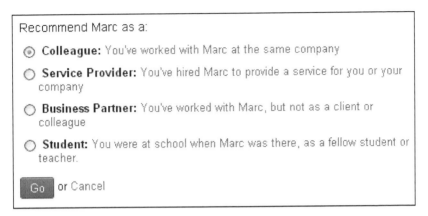

Figure 10: Recommend Contact

The next page asks me to clarify the relationship, where I choose my title and his title when we knew each other (or worked with one another). This is one of the clumsiest parts of the process for me because I've met lots of my contacts through networking events or through volunteering. I choose the closest "right answer" so I can get to the actual recommendation.

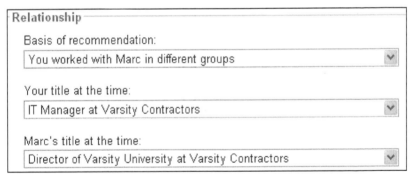

Figure 11: Clarify Relationship

The next box is where you write your recommendation. These are not long narratives. A recommendation is usually the length of a paragraph. Here's a recommendation I created for, Marc, one of my contacts (I leave a blank line after the first sentence, but LinkedIn strips that out):

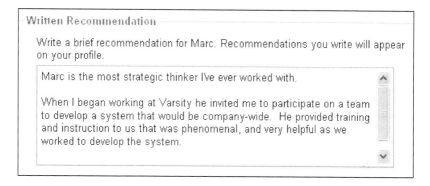

Figure 12: Written Recommendation

After you write the recommendation and hit "Send," your contact will get an email with a link to "Display" the recommendation. This will allow them to show it on their profile, like this one that I received from Janet:

This recommendation is not yet displayed on your profile.
Display Janet Meiners - Janet@AffiliateFlash.com's recommendation

Figure 13: Display Recommendation

When you click on the link, you are taken to LinkedIn where you can choose to show or hide the recommendation. I am commonly asked for a replacement from people who are working on their brand, and they ask me to rewrite it to emphasize certain characteristics. This is similar to getting a letter of recommendation that isn't quite "right" and asking the person to rewrite it.

Figure 14: Show/Hide Recommendation

One of the best ways to get a recommendation is to recommend someone. Since recommendations can be valuable endorsements used for business purposes, make sure you give the right message when you write them. Include specific details in the recommendations you give. The value of recommendations diminishes when they lack specific points.

TIP: If you want a recommendation, give a recommendation! You are more likely to get a good recommendation if you give one first.

It is okay to ask for specific wording in a LinkedIn Recommendation. When Janet was preparing my recommendation, she asked me if there was anything specific she should mention. This is an excellent time to communicate your brand to your contact, who might not understand the full scope of your current brand/offering. Take the opportunity to educate your contacts. This will make your recommendation more relevant to your current needs.

As you work on getting the right recommendations for your own profile, consider these two points:

First, some people think you can have too many recommendations. But I would never discourage someone from writing a recommendation for me. What's the right percentage of recommendations, based on the number of first-degree connections? I don't have an answer for that. However, I think you should have at least four to eight recommendations.

Second, I have seen profiles of people who were in a job search and had a high percentage of recommendations from their previous company. In these instances, I knew the people had been recently laid off from the company, and the recommendations from their previous colleagues and executives seemed fake.

Recommendations can be a powerful way to reach out to a contact and strengthen your relationship. I don't think you can really do this wrong, as long as you remember this is more of a professional network than a casual, social network. Keep the tone of recommendations similar to what you would see in a letter of recommendation; you will give a real gift to your contacts.

For an excellent primer on how to write effective recommendations, visit Naina Redhu's blog post.[14] In her post, she also describes how to write a mediocre and a least effective recommendation.

Here are my five top, um, recommendations regarding recommendations:

1. Only ask for recommendations from someone who can give you a real recommendation.
2. Only give recommendations when you have something you can honestly say about the person.
3. Give recommendations without demanding a reciprocal recommendation.
4. If you feel forced to give a recommendation, don't do it.
5. Give recommendations that are specific, speaking to professional competencies, skills, attributes, and characteristics.

---

14. http://tinyurl.com/RecExamples
    aside.in/blog/networking/2006/10/02/linkedin-recommendation-examples/

Chapter Summary

- Adding recommendations to your profile is a great way to substantiate your strengths.

- Giving recommendations is an excellent way to reach out to your contacts and strengthen individual relationships.

- It is okay to ask for recommendations, just as you might ask for a letter of recommendation from a boss.

- It is okay to ask for revisions of a recommendation, especially if the recommendation isn't in sync with your branding strategy.

- Write recommendations that are specific and professionally endorsing, not vague or general.

"It's what you do and how you leverage your LinkedIn profile that makes you who you are to your network, and be memorable for time to come."
Dan Schawbel, Managing Partner, Millennial Branding LLC; Author, *Me 2.0*; and Publisher, Personal Branding Blog, http://www.personalbrandingblog.com

"Meaningful endorsements from people you have worked with in a team or done projects for are the primary validation currency of positive attitude and achievements. It is the most important cutting edge tool for identification of these positive leadership traits. When hiring managers look at resumes, reading LinkedIn endorsements should always be the next tool used to narrow the pile."
David Dalka, Senior Marketing and Business Development Professional, http://www.daviddalka.com

---

9 Jobs

The Jobs section is LinkedIn's version of a job board. A major benefit of LinkedIn's job board is that you can see your relationship with the people who post the jobs. You can see how many degrees they are from you, how many recommendations they have, and what their title is. I love seeing postings by a hiring manager, since that is exactly who I want to network into as a job seeker. If I'm doing competitive intelligence research, even as a salesperson, I just learned the name, title, and role of someone who is in a position of power.

CALLOUT TIP: Whether you are in a job search or not, use job boards to gather intelligence about your target companies.

Showing who posted the job sets LinkedIn apart from other job boards and is the most powerful feature of LinkedIn's job board. Can you imagine if every job board gave you information helping you network into the company? That's what LinkedIn is doing, and it is awesome!

Some people get excited at the idea that many jobs posted on LinkedIn are exclusive to LinkedIn users.

In 2007, I did a search for jobs in New York City and got 415 results (job postings). In 2008, I did the same search and found 715 results. In late 2010, I did the same search and found 1,226 results. These results were based on a keyword search (in other words, I put "New York City" in the search box). A zip code search, using a New York City zip code (and a fifty mile radius) returned 2,662 results.

A similar search on Indeed.com (New York) produced over two hundred thousand jobs. A search on Linkup.com produced over twenty thousand hits. I definitely wouldn't recommend LinkedIn as your main job board, but there are two reasons I would encourage you to keep it as a tool:

First, as mentioned, LinkedIn makes it easy to network into the job opening or company. Second, some opportunities are only posted on LinkedIn.

When I click on a job posting, I can see how to network into the company. In addition to seeing who posted the job, I can see who can give me an introduction to that person. I can also see how many people in my network are at that company, which helps me figure out other ways to network into the company.

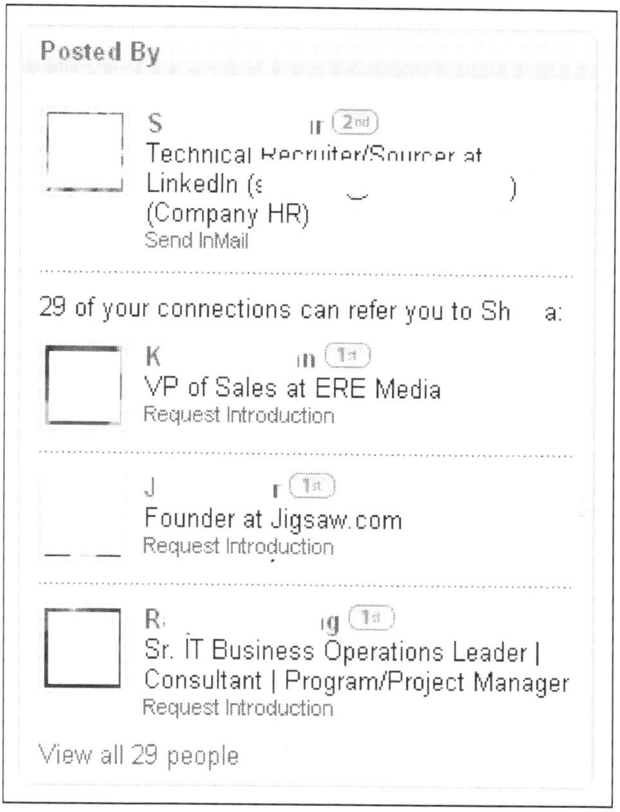

Figure 15: Job Posting Inside Connections

With tens of thousands of job boards (including corporate job boards), it is confusing to know which boards and tools you should be using. I suggest the LinkedIn job section is one of the job boards you use because of the networking information LinkedIn Jobs provides. If you "Save" the job, you can have an email alert sent to you daily, weekly, etc.

In the last edition of this book, I asked LinkedIn for some improvements. Here's a report of the requests I made:

1. **Include salary as a search criterion.** This is not in the system. It has become commonplace to not list a salary, allowing the poster to say "DOE," or "depending on experience," so I don't now if allowing you to do a salary search will ever be possible. But it would be a great enhancement for the job seeker.

2. **Allow me to save my searches.** Yes! I can now do this and have my saved searches come to my email, like most job boards. Thank you, LinkedIn!

3. **Send me email alerts of jobs based on my criteria.** Yes, as I mentioned in #2 above.

4. **Get more job postings.** They are making nominal progress, but the growth of the job postings is still slower than the growth of their overall sign-ups.

Let me share some general job board thoughts, which apply to you whether you are looking for a job or not. Job boards are an excellent place to do research on a company. If your business has competitors or wants to do research on a target company's competitors, use job boards to see what's going on.

For example, let's say you are trying to figure out what direction a hiring company, or competitive company, is going. You search for them on a job board and find they have openings for two key roles in their customer service department. This might mean they really are having customer issues (something you might have heard while interviewing others). Or, it might mean they are expanding their customer base. If the same company is hiring ten sales executives in New England, you know they are expanding in New England (where they might not have had offices before).

Job boards may not have the best record for landing a job (most stats I read say only 5 percent to 14 percent of jobs are filled because of job boards), but they are rich with competitive intelligence information! Use job boards, including the LinkedIn job board, to gather information about your target companies and figure out how to network into the right decision maker!

Chapter Summary

- LinkedIn Jobs shows you the people in your network that have some tie to any particular job, which can help you network your way into a company.

- Use LinkedIn Jobs to do competitive intelligence research, whether you are looking for a job, looking for a customer, or just checking out your competition.

10 Companies

When I wrote the first edition of this book, researching companies was limited to whatever you could find when doing a People Search. A year later, by the time I wrote the second edition, LinkedIn made significant enhancements and added a "Companies" section.

The addition of companies was significant. LinkedIn has a rich database of company information that is enhanced each time a user adds information on their profile. LinkedIn shows some of this information to users. For example, I can see where employees of my target company worked before they came to my target company, and then where they tend to go after leaving the target company.

This is great information as I gather intelligence on my target company! I might not have known, or thought of, these tangential companies, and it expands my research. There are other types of information like this that are valuable in doing company research.

Let's explore a few ways to get value out of Companies. First, you need to get to a company page.

When you click the "Companies" link on the main menu, you are taken to the Company Search page where you can search by industry, country, zip code (postal code), or by company name.

CALLOUT TIP: You should be able to list, right now, three target companies. What are they? Write them on this page!

Another way to find a company page is from a user profile. If the company name is a hyperlink, you can click on it to see the company page. For example, search on profiles for IBM. On most results, the IBM will be a hyperlink that you can click on to get to the company page.

If you aren't sure what company you want to look at, or network into, think about the type of person you want to network with, based on titles, roles, where they are located, and then look at the companies listed on their profiles.

LinkedIn users can add companies into the Companies database. You can add the company easily by clicking the "Add a Company" link, but only if you have a corporate email address (in other words, you have to have an @companyname.com address).

On the company page, you'll see a company summary at the top of the screen. This summary gives general information that you can find on Google or Yahoo! Finance. There is more information LinkedIn shares that is really rich, such as

- where company employees worked before they joined that company,

- what companies ex-employees go to after they leave the company,

- who in your network works at the company,

- new hires, promotions, or job changes,

- key company statistics,

- and more.

In addition to getting this information for your target company, you can also figure out how to network into the company! If you are in sales, looking for partners, or are a job seeker and you have relevant contacts, you should see more ways to network into the company than if you didn't have any relevant contacts. This is another compelling reason to increase your network with relevant contacts.

LinkedIn is beta testing a "Company Group" feature, which is open only to those currently employed by certain companies. I don't have much information about this but hope LinkedIn enhances it, as it could provide a great networking opportunity for company employees. There are many groups for company alumni. Click on "Groups" and search for companies you used to work at to see if they have an alumni group.

One of the more recent enhancements to Companies is the ability to "follow" a company. Once I follow a company, I can see from the Settings page new information about the company, either via Network Updates or email. You can choose to get updates when employees leave or join the company or when they are promoted. You can get new job postings sent to you as well as any Company Profile updates. It's a great way to keep news about the company in front of you.

Chapter Summary

- Companies have data from various sources to provide you a rich tool for research.

- Use Company Pages to do competitive intelligence research on target companies, or competitors of target companies.

- Find target companies by using the Company Search page or by finding companies whose names are hyperlinked from profiles.

- Learn about your target company and determine how you could network your way into the company or strengthen your contacts in that company, based on who is already in your network.

"The Company Profile pages on LinkedIn are GOLD for anyone using LinkedIn to research companies, individuals or positions within a company—and for recruiters searching for candidates! The Company Profile page contains a wealth of information on the company via Business Week and employment data compiled from the information current and former employees have entered in LinkedIn. This market intelligence is fantastic, allowing you to see hiring and promotional activity within the company, as well as quick links to additional information to research competitors, similar companies, etc. As a recruiter and avid networker, this is one of my favorite features within LinkedIn, and it can be used for business development and job search purposes in the same way."

Jennifer McClure, Executive Recruiter/Executive Coach; and President, Unbridled Talent, LLC,
http://unbridledtalent.com/blog

11 LinkedIn Groups

The value of LinkedIn Groups has really increased, even since the last edition of this book. If you only choose to do a few proactive things on LinkedIn, Groups needs your attention.

At the time of this writing, there were over 747,000 LinkedIn Groups. That equals 650,000 more groups than there were just two years earlier. That means almost one thousand new groups are created each day! Don't worry, you don't need to know about most of them—but there are a *few* you should know about!

There are many kinds of groups, including groups for alumni associations, professional affiliations, industry interests, special interests, company alumni groups, and generational groups (GenY, Baby Boomers, etc.).

To join a group, simply click the "Join Group" button on the Group Page. Some groups let you in automatically, while others have an approval process. For example, you can't join the Yale University Alumni group until they verify that you are an alumnus of Yale. If you are not automatically approved, you can send a message to the group administrator asking for approval to join the group. I regularly get this type of "please let me join your group" email for my JibberJobber Career Management group (search "Jibber-Jobber" in the Groups Directory.

Once you become a group member, you have access to browse through a list of members as well as search by keywords. If you have time, look for key prospects you want to network into.

One of the biggest benefits of joining a group is that you can message other group members, whether they are connected to you or not. This could be the most important thing for you, even if you don't do anything with Groups. The ability to message members is so powerful, I cannot overstate it. Imagine you only have five first-degree connections. You find a group relevant to you with one thousand members. You now have direct reach to all one thousand of the members of that group. Multiple that by the fifty groups you can join—and imagine the massive reach to relevant contacts you can have just by joining relevant groups!

In my business, I've been prospecting university career centers, since I developed a very cool product for them. I found five groups that career center professionals join, and I now have reach to thousands of career center professionals, without even connecting to them. When I find a profile of a career center professional I usually can communicate with them because they are usual a member of one of the groups. The alternatives include upgrading or asking for an Introduction, both of which I don't want to do.

Another significant feature in Groups is "Discussions." Discussions gives you the ability to post a question or thought to group members, or respond to their posted questions or thoughts. This presents a significant opportunity to brand yourself as a thought leader or subject matter expert, *if you do it right*. There is a lot of spam and inappropriate content in group discussions. I encourage you to keep your conversation highly value-added. You don't want to be known as a spammer.

The ability to create discussions is almost like having a blog with a built-in audience. Don't disrespect the audience by posting too much or stuff that is irrelevant. Also, make sure anything you put up (a new discussion, or a comment on a discussion) is on-brand for you.

When you join a group there are group settings you can adjust. One setting is the option to show the group logo on your profile. I can show my support for, and affinity to, certain causes, associations, clubs, companies, and schools just by showing the logo on my profile.

You don't have to show all of your groups on your profile. Show only the groups that support the brand message you've defined. Only show enough groups that get the point across. You can still have full membership in a group without showing the logo on your profile. You don't want to add more noise on your profile by adding showing too many groups.

Look for all of the relevant groups you can find and join them! Whether you have a proactive strategy and participate in group discussions, or you have a passive strategy and just join to be in the group, you can't go wrong.

Owning Your Own Group

There are a lot of groups—maybe even one million groups by the time you are reading this! Why should you start your own group? There are a few really good reasons.

I started a group just so I could test out the features and see what it was like to be an owner. As a group owner I can:

- approve new members, or remove people who spam the group, or abuse their membership;

- moderate all discussions (not that I have to, but I can), and essentially have the last word; and

- send an "Announcement," which is like a newsletter, to all group members.

Let's talk about starting your own group. Click on "Groups" to get to your "My Groups" page, and then click on the "Create a Group" tab. You'll need to have a name and description for your new group and, optionally, a logo.

Once your group is created, your job is to let people know about the group! There is no easy "tell everyone about my group" button. Let all of your network contacts, clients, and prospects know about the group through traditional methods, including your email signature, your website or blog, in newsletters, in comments you leave on other blogs, in group discussions, etc. Also use non-LinkedIn communication tools to let others know about your group.

CAUTION: Be careful how you do this on a seemingly competitive group. No one (especially the owner of the other group) wants to read spam messages.

When you group up, make sure you have the right keywords in the description and summary so people can find it when they search for what your group is about. What are the keywords that best describe your group? What are keywords your target audience would use to search for your group? List them out and then incorporate them into the group. Also, visit competitive groups to see what keywords they use to get other ideas.

One benefit of owning a group is the viral marketing your brand/company/expertise get. When group members put your logo on their profiles, they are advertising you. Many of my JibberJobber Career Management members have the JibberJobber logo on their profile, which helps my branding.

Another way your group spreads virally is when someone joins the group, or contributes to a discussion. Their connections can see, on the homepage, their group's activity. The viewer will see the title of the Group Discussion and can click a link to learn more.

As the group administrator, don't be shy to develop your own rules and boundaries. When someone is abusing the group to send promotional messages, kick them out. Jump into the discussions when you have time and keep it alive. Also, take advantage of the Announcements section where you can send an email message to each member's email box. This essentially turns your group into an opt-in newsletter. Just don't abuse it.

Let's wrap this up. As a user of LinkedIn, I encourage you to:

- Seek out and join groups where your peers and target audience are. You can join up to fifty groups, but don't show all fifty group logos on your profile, as it will just clutter your profile. Join as many groups as you can.

- Go into your groups and browse the list of members. Perform from inside the group to find target prospects.

- Reach out to group members by sending individual messages. Make sure you have a purpose to reach out to them and clearly communicate your purpose.

- Find discussions you can join or start. Pay attention to what you say, and how you say it, and avoid any netiquette missteps.

If you are interested in owning a group, I encourage you to:

- Go to Groups, click on the "Create a Group" tab, and get started on your own group!

- Market your group in as many places as you can. This includes your email signature, on blogs and websites, in comments you leave on other blogs, and in newsletters.

- Figure out how open your group is. If it's an alumni group, you'll want to limit access to alumni of the school or company. If it's a special interest group, you may be more open. Determine how open you'll be based on how much administration you can do (do you want to have to check out each applicant before you approve them?).

- Spend time in your Groups Discussions to keep out spam and weird messaging. Encourage good conversations.

- Communicate to group members by initiating discussions and sending out announcements.

Participating in groups has added value to my LinkedIn experience. My group has over two thousand members. Other groups have hundreds of thousands of members. Don't discount smaller groups, as your voice will still be heard by others and not get lost in the noise of a larger group.

Can you see how groups could be an essential part of your LinkedIn strategy?

Chapter Summary

- Join groups where your peers or target audience are. Consider adding certain group logos to your profile.

- Communicate with other group members using Groups Discussions, but always keep it on-brand.

- Look for group members to see who should be in your network, or who you should initiate a conversation with.

- Consider starting your own group for your company, profession, industry, passion, and local geographic area.

"Administering a LinkedIn group is a terrific way to subtly raise awareness of yourself. It's a fairly simple process to start a group. Perhaps the most difficult part is designing your group's logo. I started and manage three groups, one for my high school and two for my geographic area, and I can see that most people who join them review my LinkedIn profile. That puts me in front of many people who may never have been aware of me otherwise. It also gives me the opportunity to welcome them and reach out to them as potential connections. It can work for you too, so consider what need is unfulfilled and start your own LinkedIn group."

Christine Pilch, Co-owner, Grow My Company; and Coauthor of *Understanding Brand Strategies: The Professional Service Firm's Guide to Growth*, http://www.growmyco.com

"Everyone knows that it's a good thing to network, but most people don't do it unless they need something. Building a LinkedIn network is a little like that. You really won't appreciate just how valuable your network has become until you need it. My advice is to start now! Take Jason's recommendations to heart and get going. When the time comes, you will be glad that you did."

Simon Meth, Corporate Recruiter & Career Counselor, Kettlehut and Meth, http://www.martinandsimon.com; **and Blogger,** http://community.ere.net/blogs/sittingxlegged/

12 LinkedIn Answers

LinkedIn Answers is one of my favorite features on LinkedIn. Why? I use LinkedIn (and other social tools) for three things:

- to reinforce and share my brand,

- to grow my network with relevant contacts, and

- to nurture individual relationships.

LinkedIn Answers lets me do all three of these things. Let's explore how. There are two aspects of LinkedIn Answers: asking questions and answering questions.

When you ask a question in Answers, you simply post a question that is viewable by anyone on LinkedIn. You can also send the question, via email, to up to two hundred of your first-degree contacts. Questions posted are extremely diverse, ranging from knowledge-based issues ("I have a technical question...") to help in finding a job ("How do I network into that company?") or resources ("Do you know where I can find the most updated list of...?"). In general, it is inappropriate to ask the type of social questions you might find on Facebook or Twitter.

You are limited in the number of questions you can ask on LinkedIn Answers. Historically, the limit has been ten questions per calendar month.

I encourage you to ask questions based on how many contacts you have. Remember, you can have LinkedIn email the question to up to two hundred of your first-degree contacts.

If you have less than two hundred contacts, submit a question about once a month. You don't want your contacts to feel like you are spamming them.

If you have less than four hundred clients, ask a question every other week. Send the first question to the first half of your network and the second question to the other half of your network.

You do this by last name, grouped alphabetically. So, the first question would go to all of the A–Ms in your network and the second question would go to the N–Zs in your network.

If you have eight hundred or more first-degree contacts, ask a question once a week. Each week, choose a different block of two hundred contacts to get the question via email.

Here's the key: I assume no one will see my question on LinkedIn. Why ask the question, then? Because I want to use LinkedIn to get my question (and my brand and my messaging) in the email inbox of my contacts.

This is really powerful. LinkedIn gives you a tool to get your message in the inbox of hundreds of people, at no cost to you. Ask questions in LinkedIn Answers to

1. share your brand,
2. grow your network, and
3. nurture relationships!

TIP: Ask a question at least once a month. This helps you intelligently probe your contacts in a creative way.

If you don't invite up to two hundred people from your first-degree network to get the question via email, you have missed the point of LinkedIn Answers. Based on my experience, you are more likely to get more responses if you choose two hundred contacts than if you just let each person randomly find your question on LinkedIn (which is not as easy as it sounds).

Even if no one responds to my question on LinkedIn, I still want my brand and message to get in front of my network.

Each question is open for a certain period of time, currently eight days. During that time, anyone on LinkedIn can answer the question. At the end of the eighth day, the question closes and no one can answer it. The question and all answers are always available for anyone to see.

If you have a good question that received good answers, consider opening the question again. Simply go to the question page and "Reopen" the question. I asked a question, got great answers, and then reopened it for another period. When I did this, I chose one hundred different first-degree contacts to get the question emailed to them.

During the eight days your question is open you can clarify or close your question. If you need to clarify the question, append your clarification at the end of the original question instead of changing the entire question. If you change the original question, some of the answers might not make sense (since they answered the previous question). I've also seen people use the "Clarify" link to follow up and respond to answers, express appreciation, etc.

What a terrific source of knowledge LinkedIn Answers is! When someone answers a question, you learn more about them, their passion for the topic, perhaps the depth of their expertise. With one click you are led to their profile where you can see if they might be a relevant contact.

Once the question is closed, you should choose the "good" and "best" answers. You don't have to, but if you do you can further the relationship with each person. You get a chance to put your brand in front of them again.

When you assign a "best" on someone's response you build the credibility of the person who answered. When they answer other questions it shows how many best answers they have gotten.

If you ask a question that others consider inappropriate (like asking a question that is a job posting, or pitching your product or company) it may get flagged. This is a peer monitoring mechanism to help eliminate spam in Answers. Be careful how you word questions (and answers to other people's questions)! You don't want to brand yourself as a spammer in Answers.

Some LinkedIn users regularly answer over 150 questions each week. You can see these people on the Answers page, below the main questions. They are called "This Week's Top Experts," even though answering a lot of questions doesn't necessarily make you an "expert." At the time of this writing, the person with the most answers this week had answered 463 questions.

If you use an RSS reader, you can get new questions in certain categories that others post on LinkedIn Answers. Simply go to the Answers page and scroll down to see the list of categories (under "Browse"). When you click on a category you'll see an RSS icon that lets you follow all new questions. This helps you see new questions without logging in to LinkedIn and browsing through the Answers page.

You should have a LinkedIn Answers strategy. Do not let a month go by without asking a question on LinkedIn Answers! Your questions should respect your network and reinforce your brand. Each question is an opportunity for you to put your messaging in front of your contacts (and people outside of your network). Here are two example questions:

- What are the top project management failures this year? (This question reinforces your interest, or expertise, in project management.)

- What are the top three challenges for supply chain professionals for the next year? (Doesn't this help me understand your interest or thought leadership in the supply chain space?)

You respect your network when you ask real questions that allow them to share their expertise. You reinforce your brand by the type and topic of each question. Don't miss the opportunity to use Answers as a powerful networking tool! Avoid pitching your offerings, or coming across as too self-promotional.

Some people disagree, but I think it's appropriate to leave links in your questions or answers. Exercise discretion and only leave links that add value to your point(s). You can link to a blog post you wrote, to a news article, or to someone else's blog post. The objective is to be a value-added contributor. Share as much pertinent knowledge as you can to be seen as the subject matter expert that you are.

TIP: Try to answer a question **at least** once a month, which will enhance your visibility in LinkedIn.

Chapter Summary

- Participating in Answers is an excellent way to make new contacts as well as reinforce your personal brand.

- Make sure you participate and add value to questions and answers.

- Your Answers strategy can include just asking questions, just answering questions, or a combination.

"When you ask a question on the Answers forum, take the time to thank each person who tried to help. Then remember to close and rate the question. People have taken the time to help you; it is simple common courtesy to thank them."
Sheilah Etheridge, Owner, SME Management,
http://www.linkedin.com/in/smemanagement

"Use the questions and Answers features to start conversations, create community, and position yourself as a subject matter expert. By answering questions, you are simultaneously endorsing your candidacy and expertise."
Barbara Safani, Owner, Career Solvers,
http://www.careersolvers.com

Part III
Wrapping It Up

There's more to getting value out of LinkedIn than the tools and the interface on their website. Part III discusses concepts for using LinkedIn in a personal branding strategy, warns you about shady practices that you might encounter, discusses netiquette, talks about complementary resources, and leaves you with some final thoughts.

13 LinkedIn for Personal Branding

I'm frequently asked how a company can make money with LinkedIn. Indeed, a company can, and should, use LinkedIn for various reasons, including corporate branding and increased sales.

In general, though, LinkedIn is a personal tool. The user profile is for one individual (not a company, for that you would use a Company Profile) and usually shows the individual's work history. I think your account is yours, not your employer's; however, that is being debated in the courtroom as recruiting firms and other companies claim the network (connecting) was done at work and it is owned by the employer.

Your LinkedIn profile is not a corporate advertising channel, although you can put language on your profile that explains what your company does in a way to attract new customers.

Developing and sharing your personal brand is not as easy as throwing up a profile. Understand what your personal brand currently is (everyone has a brand, whether they know it or not) and define what you think it should be. Figure out how you can strengthen it, online and offline.

Since this is not a book on personal branding, I won't cover all the ins and outs of who you are or who you should be and explain all the ways to share and reinforce your brand. I will share some ideas on personal branding specific to LinkedIn. If you are interested in personal branding, I highly suggest the book *Career Distinction* by William Arruda and Kirsten Dixson. Another popular personal branding book is *Me 2.0* by Dan Schawbel. Two excellent, no-cost personal branding resources are Dan Shawbel's popular Personal Branding Blog, http://www.personalbrandingblog.com, and the site A Brand You World, http://www.personalbrandingevent.com, which has hours of recordings from its 2007 Global Telesummit.

Here are my recommendations for you, as you work on your personal brand using LinkedIn:

First, make sure you are showing the right information on your Public Profile. LinkedIn allows you to view your profile as others would see it. You should log out and see what your profile looks to people who are not logged in (people who find your profile from a Google search, or who get a link to your profile).

Is your profile showing enough information? Is it showing too much information (and muddying up your brand message)? Too many profiles show too little information, which makes me spend time poking around the Internet for more information on you. Unfortunately, I don't have the time (neither does your target prospect), and I move on to someone who has presented me with enough information to make a decision. Make it easy for people to learn about you without going somewhere else!

Second, make your profile look intentional and take advantage of subtle branding opportunities. Change your Public Profile URL from the default assigned value (which looks like gibberish) to something more descriptive of who you are. For example, you can see that my Public Profile ends in "jasonalba" instead of something like "/1/234/698." If you see a link to my profile, you know exactly where you are going:

Public Profile	http://www.linkedin.com/in/jasonalba Edit

Figure 16: Edit Profile

To change yours go to the Profile page and click the "Edit" link. Don't put something regretful, like "hairy legs."

Third, share the link to your LinkedIn profile. The easiest, most effective and most viral way of sharing this is to put it in your email signature. Every time someone reads your email signature, they'll know how to learn more about you. Each time your emails get forwarded, more people will see your profile URL. Another way to share your profile is to put the "View My LinkedIn Profile" image on your personal website or blog.

NOTE: I DO NOT DO THIS. Why not? I consider my LinkedIn profile a channel, to get you to my other sites, which are the destinations. If you don't have a website or blog or other destination, use your LinkedIn profile. But think about where you want people to end up and send them straight there.

I attended a presentation for job seekers about "regaining your identity." The speaker talked about using resumes in networking events. He said that employed professionals don't pass out resumes, they trade business cards. If you get a resume from someone, you have a preconceived idea about what they want (a job). However, if they give you a business card, you are just two peers, two professionals, exchanging contact information.

This concept applies to your LinkedIn profile. If you send a link with your resume with every email you send you'll likely get into trouble with your employer, or your send a message that you are needy. Sending a resume says, "I'm looking; if you can help, then let me know." However, sending a link to your LinkedIn profile says "Here's my professional profile, check it out."

Of course, your LinkedIn profile can have the same information I'll find on your resume, with the same quantifying statements. But just pointing someone to your online profile sends a different message.

Finally, if you leave a comment on a blog, online newspaper, forum or some other website, you have two opportunities to leave a link to your website. If you don't have your own website or blog, put in the link to your LinkedIn profile URL. The first place to put a URL is in the "website" box. Usually, when you leave a comment on a blog post they ask for your name, email address, and website.

The second place to leave the URL is at the end of your comment, much like an email signature. Make it easy for the reader to click a link to learn more about you!

Chapter Summary

- Use LinkedIn to strengthen your personal brand and take advantage of LinkedIn's search engine optimization.

- Your LinkedIn profile is kind of like a resume, without many of the limitations (length) and negative connotations ("I'm a job seeker, help me!") that a resume might have.

- Use your Public Profile URL as your "website" address when you comment on blogs, unless you have a blog or other website you want people to go to.

14 Shady Practices

With as many users as LinkedIn has, you are almost guaranteed to run into people who don't abide by common netiquette or the LinkedIn User Agreement (there is a link at the bottom of each page to the User Agreement).

Here are some actual practices I've seen that are considered to be "shady":

Email in name field. This is a very simple, almost nonoffensive deed, but I mention it because it is against the terms set forth in the LinkedIn User "Dos" and Don'ts" section of the User Agreement:[15] in which the user agrees not to post "to a content field content that is not intended for such field." Example: putting an address in a name or title field. This may be the most common offense, to the point where it looks like "everyone is doing it." Maybe everyone is doing it, but it is still against the rules.

Tollboothing. This is when people charge you to have access to their LinkedIn connections. You can do a search on their network, but if you want an introduction or some other kind of endorsed

---

15. http://tinyurl.com/UserAgreement
    linkedin.com/static?key=user_agreement

communication to one of their contacts, they charge you. The logic behind this is, "I put a lot of time and effort into building my network, and this is my livelihood, so you should expect to have to pay for my services." I don't agree with this behavior. While there are people who agree with it, they are in a small minority. Doing something like this might hinder your networking and LinkedIn strategy.

Lying. One of the best ways to be found by companies that hire Yale graduates is to graduate from Yale, right? What if you could just put Yale as one of your schools that you went to? Well, you can. You can lie about places that you worked and where you went to school. You can actually lie about anything on your profile. This is one reason why recommendations carry weight, because you can't edit what someone else writes about you. Some people put false information on their LinkedIn profile to increase their chances of being found by recruiters or hiring managers.

Vincent Wright, networking enthusiast and Chief Branding Officer of Brandergy, found a profile where the person had supposedly worked for fifteen years in one hundred different companies. Vincent called him the 1,500-year-old man (100 companies x 15 years at each one = 1,500 years of work experience!). Vincent is a professional recruiter—and professional recruiters can see this type of deception from a mile away. Make sure your profile is clean and verifiable—just as your resume should be.

What should you do if someone contacts you because of a commonality with a school or company listed on your account? On an email listserv, Scott Allen recommended the following:

> If someone sends you an invitation saying they're a former colleague or classmate, at least take a look at the dates and other entries on their profile and make sure they jibe. If they don't have a full position and description listed, that's a yellow flag.

It's up to you to make sure the person you connect with, or respond to, is who he says he is. I don't expect LinkedIn to verify the information. When I get a communication from someone on LinkedIn that is fraudulent, I forward the message to the customer support email address with hopes their account gets deleted.

Using LinkedIn Answers to cloak inappropriate questions. The LinkedIn community has been clear they don't want Answers to turn into a spam-laden area, generating useless requests for information and sending out numerous emails. Try a few questions that are borderline and you'll probably have some people respond that it was inappropriate. Answers has a purpose. Abusing it not only clutters and diminishes the value of the system but also shows you don't respect others.

Fishing for fake recommendations. You can't make up recommendations for your profile, and creating fake profiles to endorse your main profile is too much work. Some people ask for recommendations without knowing you. The shady part is when someone you don't really know asks for a recommendation. It usually comes in the form of "If you recommend me, I'll recommend you." I used to respond with something like this: "I don't really feel like I know you well enough to give you a recommendation—sorry. I'd like to get to know you better before I do that." I never got a response back, and now I don't even bother sending that message. I think some people send requests for recommendations to their entire network, so I just ignore them if I don't know them well enough.

Light-linking. This refers to linking with anyone who asks for a connection (this is something a LION would do). I said I'm not going to tell you what your connection philosophy should be, but if all we have on LinkedIn is connections without relationships, we do minimize the value LinkedIn could bring us.

The value of LinkedIn is diminished when connections are made with people you don't know well enough to endorse, recommend, or pass along to another network contact. You can argue that light-linking is okay since it

a. expands your network and reach, and
b. increases the chances that you'll be found because your network gets bigger and more diverse.

Fishing for email addresses. Because so many people violate the rule about putting email addresses in the name field, it's easy to find email addresses to put into a list. I'm not suggesting that *you* do anything shady, but here's how easy it is. Just search on ".com" in the search box and you'll find plenty of email addresses to add to your list. It's that easy to find email addresses to spam!

These are not all of the shady practices—the key is to be cautious with the information you get and be careful with the messages you send.

Chapter Summary

- People will do things wrong, sometimes unknowingly. Beware of whom you interact with, and how you interact with them.

- If you violate LinkedIn policy, you risk having your account suspended, or being seen as someone that does not respect "the rules."

15 On Netiquette

With all the new social interfaces on LinkedIn, there are plenty of opportunities to make etiquette mistakes with your contacts. Netiquette refers to etiquette, or acceptable social behavior, online (hence, the "net").

It's critical to understand some basic rules of Internet etiquette. Not understanding can have consequences as simple as people ignoring you and as complex as getting kicked out of various forums and websites. In addition to having an online presence and developing your personal brand, you should stay within the boundaries of your professional brand. I assume some component of your brand is awareness, professional consideration, and respect. Even if you are "aggressive," you should still have respect for the rules of interaction. Here are some basics rules to consider:

1. **Be nice. Be concise.** Always! Aren't we in the habit of *skimming* more than reading? We're faced with massive amount of information from all sides: email, social websites, Internet searches, blogs, news sites… the list doesn't seem to end! Have a

concise message with a nice tone and I bet you'll have your message read more often. Write a long message and I'll save it for later, when I'm not busy. Know when that is? Never.

2. **Avoid sarcasm in your online communications.** Maybe it's because we're more inclined to skim messages, but it's often said we should be very careful with funny jokes and sarcasm in the written word. I frequently hit the backspace button when writing many of my emails because something that sounded good while I wrote it could be interpreted too many ways. Furthermore, jokes or sarcasm could be a distraction from my real message and just add noise, which is an unintended result.

3. **Assume good intentions when you are reading someone else's communication.** Scott Allen advises people to "presume good intent" when reading an email that might come across as negative, harsh, or inappropriate. Scott is right! You know you should be careful when sending messages, but some of your contact might not have gotten the memo on netiquette. Give them the benefit of the doubt!

4. **Don't chastise or preach.** I've seen too many online discussions go from bad to worse when one person chastises or preaches, advocating the "right" way to do something or think about something. What starts off as a general admonition or knowledge-sharing email easily turns into personal jabs and accusations. I've found it's much more effective to just bow out of the discussion and move on to something else. Everyone involved will appreciate the thread dying down, and you'll look more mature, wiser, or just smarter for not pursuing an online fistfight.

5. **Consider cultural differences when reading or writing anything.** Every list I'm on has a good representation of people from all walks of life, and from different countries. There are people whose native tongue is not the same as everyone else on the list. There are people who are right out of school (or still students) and others who are at the end of their career. There are entrepreneurs, executives, artists, rich people, and poor folks. These differences affect the messages and the culture of the online forum and can lead to misunderstandings.

6. **Know when to take it off-list.** Lots of topics discussed online are interesting and appropriate. Sometimes they are only interesting and appropriate to the two people participating in the discussion. If this is the case, take the discussion off-list, which means to do it outside for the general group. If the discussion starts off slightly off-topic and it gets deeper and more technical, take it off-list. If the discussion is between you and one other person, and adds no value to anyone else, take it off-list. If the discussion is completely off-topic, or personal, take it off-list. If you are just saying "thank you," or "happy birthday" or something like that, take it off-list.

There are always exceptions to the rules (it seems many people don't know the rules *or* the exceptions). Remember this: everything you write, whether it's a comment, an email, an instant message, or anything, may come back to haunt you. I like the rule of thumb to write everything as if it will be printed on the front page of a major newspaper, with full credit back to me!

Chapter Summary

- Learning the rules of netiquette will help you maintain the proper relationships as you communicate with people online.

- Understanding netiquette mistakes may help you be more understanding or patient with people who don't understand them.

16 Complementary Tools and Resources

I have grown to love and appreciate real networking. I've found LinkedIn to be an important, valuable, part of my networking strategy. There are other network tools I recommend that complement LinkedIn. This chapter lists those tools.

Many people use LinkedIn to develop an online personal branding strategy. This means they develop a profile much like they would develop a resume, trying to optimize it so people will

a. find it when searching for keywords (such as "project manager"), and/or

b. be interested in working with, hiring, or finding out more about the person.

If you are interested in developing an online presence there are other ways to do it. A Public Profile on LinkedIn should be one part of a multi-faceted strategy, not your entire strategy. Here's a quick list of excellent (and free) ways to enhance your online presence, and even get your name on the front page of search engine results:

- Find other sites that allow you to have a public profile, such as Jobster.com, Emurse.com, JobSpice.com, VisualCV.com, and ZoomInfo.com. There are hundreds of sites that allow you to have a profile that search engines can access.

- Set up a blog that allows you to brand yourself and quantify your professional **breadth** and **depth** in a very powerful way. If you are open to this idea, consider Twitter, which is frequently called a "microblog."

- Comment on other blogs to establish your online presence and footprint, pointing back to a central place (either your blog or one of your online profiles (like your LinkedIn Public Profile).

- Develop a Squidoo lens where you can list things such as your online profiles, favorite books, and blog feeds. The goal is to help people understand your professional interests and abilities and to be found when people search for certain keywords.

- Write articles and post them for free online, or volunteer to write a column for a magazine or newspaper.

All of these tools can complement one another. On your LinkedIn profile, you can list your other websites and profiles. On your blog, you can list your LinkedIn profile and link to your articles and other online tools. This allows someone who finds one of your profiles to visit other pages where they might learn more about you because of what you have there or how it is presented.

Having a presence and strategy in various environments might help you connect with others in a place where they feel comfortable. For example, your target audience might be on Twitter or on Facebook but not on other sites. Or maybe they aren't on any of these social sites, and you are most likely to be found when they use a search engine. Having multiple sites can help search engines find you. Enter "SEO" in your favorite search engine to learn more.

Let's talk about relationship management. There are hundreds of relationship management tools available to help manage your relationships. Recruiters use a relationship tool to track their job candidates (they call it an applicant tracking system). They make notes on candidates and create log entries and action items to help them as they try to fill events. Salespeople have a customer relationship management tool, which is similar. They use their relationship tool to help manage information about prospects and clients. Why shouldn't you have something as helpful to manage your relationships?

I'm a strong advocate of using a relationship management tool because that's where I've concentrated my efforts since 2006, shortly after I got laid off. I designed a system for the average "nonsales" person. Initially it was designed for the job seeker, but now I want a tool for people who aren't in transition, too. So JibberJobber.com has become my relationship management tool. LinkedIn is not a relationship management tool, but it is a great complement to a relationship management tool.

The relationship management tool you use could be as rudimentary as an Excel spreadsheet (good luck with that), as complex as a salesperson's contact relationship management suite (which might require a week of training to use), or as simple and common as Microsoft Outlook's Contacts section (which can be good enough but lacks some valuable features).

Industrial-level tools, such as the salesperson's ACT!, GoldMine®, or Salesforce.com are quite common. JibberJobber.com was designed to manage your personal relationships in a career management context. Anyone interested in "climbing the ladder," creating "job security," or developing and nurturing a professional network would find JibberJobber.com to be useful. No matter what you use, you should use something in addition to LinkedIn.

Other social networking sites have made considerable progress offering professional networking opportunities. Currently, Twitter and Facebook are regarded as tools with which a professional can network, in addition to LinkedIn. Facebook lacks the degrees of separation power, as well as a strong search interface. If they were to enhance those two features, they would present a considerable threat to what LinkedIn offers, since they have more than 600 percent the number of sign-ups. Twitter is powerful, but it has a very focused, simple feature set and lacks a lot of features that LinkedIn offers. It is definitely a powerful tool, though.

I don't advocate the use of one over the other since many might be perfect complements. When I consider a networking tool, I ask a few questions. Will this tool help me:

- Expand my network within my country?

- Expand my network internationally? There are other sites that I might need to look at.

- Voice my opinion and develop my personal brand with a targeted audience? Yahoo! Groups and Google Groups provide a great forum for this.

- Learn from other like-minded professionals? Again, Yahoo! Groups and Google Groups might provide a great environment to do this. One of my favorite examples can be found at Young-PRPros.com.

There are a lot of resources to keep up with what's going on at LinkedIn. LinkedIn's own blog and learning center have matured quite a bit and can provide you with information on new features. Here are some other resources:

- The **ImOnLinkedInNowWhat.com** blog—I started this blog to serve as a supplement to this book. It has become a great resource where I can flesh out ideas from the book, talk about current issues, and respond to readers' questions. You can sign up to get the posts by email or RSS, or just visit and search for topics of interest.

- The **Linked Intelligence** blog[16]—Scott Allen's blog which has some great posts explaining techniques and strategies to help us understand how we can get more value out of LinkedIn.

---

16. http://www.linkedintelligence.com

- Guy Kawasaki's **"Ten Ways to Use LinkedIn"** blog post[17]—This is an old but relevant list of eleven, actually, ways to use LinkedIn. Reading these ideas helps you know what LinkedIn thinks is important with LinkedIn. Make sure to read the comments for great ideas from LinkedIn users.

- Guy Kawasaki's **"LinkedIn Profile Extreme Makeover"** blog post[18]—Guy received preferred treatment from LinkedIn as they guided him through improving his own profile (which he had been neglecting). This post points out the specific changes that LinkedIn recommends to his old profile. Again, the comments have valuable insights.

- The **LinkedIn User Agreement** page[19]—If you are serious about using LinkedIn, you should read this at least once (don't worry, it is short and fairly easy to read). There are a number of violations that might cause LinkedIn to freeze your account. Get familiar with the rules and philosophies behind the rules and you should be okay.

- The official **LinkedIn Blog**[20]—This blog has really matured since it first started, and has excellent information and news for LinkedIn users and fans. There are many authors of this blog, so you get a good mix of information, including recent releases, LinkedIn best practices, etc.

- The **LinkedIn Users Manual** blog[21]—Peter Nguyen has good ideas on making money with LinkedIn, selling knowledge, etc. I don't subscribe to any "get rich quick" methodology and haven't followed his stuff a lot, but he's a proficient blogger and might have some great ideas.

---

17. http://tinyurl.com/LinkedIn10
    blog.guykawasaki.com/2007/01/ten_ways_to_use.html
18. http://tinyurl.com/KawasakiMakeover
    blog.guykawasaki.com/2007/01/linkedin_profil.html
19. http://tinyurl.com/UserAgreement
    linkedin.com/static?key=user_agreement&trk=ftr_useragre
20. http://blog.linkedin.com
21. http://tinyurl.com/LinkedInManual
    linkedinusermanual.blogspot.com/

- Deb Dib's article, **"LinkedIn—What It Is and Why You Need to Be On It"**[22]—this is an excellent article written for executives in career transition. There are eight links to very compelling LinkedIn profiles that you must check out as you optimize your own profile.

- The *LinkedIn Personal Trainer* is a book by Steve Tylock. I wish I could say I wrote the first book on LinkedIn, but Steve got his book out shortly before mine came out. Steve's has a bunch of worksheets you can use as you get your LinkedIn strategy up and running. You can find more information at http://www.linkedInpersonaltrainer.com. There are now a lot of books on LinkedIn available. Simply go to Amazon.com and search for LinkedIn.

- **GetSatisfaction.com** is where you can see current issues and make complaints about LinkedIn (and other companies). Frequently I'm asked how to resolve something, make recommendations to LinkedIn, etc. I have had very poor results trying to get replies from LinkedIn employees. But I've seen them monitor the discussions and complaints at Get Satisfaction. If you have something that isn't getting any attention from customer service, you could try posting it at http://www.getsatisfaction.com/linkedin.

In addition to these resources, do not underestimate a solid networking book to learn some networking basics. Online networking and offline networking have one key thing in common—it's all about relationships. Here are some great networking books I recommend:

- *Never Eat Alone*[23] by Keith Ferrazzi—I read this book when I thought networking was all about desperate people schmoozing and passing business cards for self-gain. It really changed my perspective on what networking is and how to do it, and I strongly, strongly recommend it to anyone who asks about networking.

- *Some Assembly Required* and *The ABC's of Networking* by Thom Singer[24]—Thom's books are great resources with hundreds of practical, right-now relationship building tips. I've found his

---

22. http://tinyurl.com/ExecLinkedIn
    job-hunt.org/executive-job-search/linkedin-for-executives.shtml
23. http://www.keithferrazzi.com
24. http://thomsinger.com/

writings to be especially applicable in the corporate environment with a lot of examples on how he enriches customer and prospect relationships. Thom's newer books include *Some Assembly Required for Women* and *Some Assembly Required: LinkedIn* ("how to make, grow, and keep business relationships using online services such as LinkedIn and others").

- *Dig Your Well Before You're Thirsty*[25] by Harvey Mackay—This has been a staple of networking books for a long time. Harvey Mackay has written a number of best-seller books on networking and career management and is an authority in this space.

- Jeffrey Gitomer's *Little Black Book of Connections: 6.5 Assets For Networking Your Way to RICH Relationships*[26]—I got this book as a gift from the guy who came up with the name JibberJobber, and it's a gift I cherish. I've been asked by multiple people to include this book as a recommendation.

Use Google Blogs search, http://blogsearch.google.com/, to see what the current blog buzz is about LinkedIn. I've listed some of my favorite resources but I'm sure there are other gems out there I haven't come across yet.

Chapter Summary

- Tools to complement LinkedIn include CRM software (mine is JibberJobber.com), discussion forums and other online profile/networking/social websites.

- Resources to complement this book include blogs, websites, and other books.

- LinkedIn should be only one facet of your online social strategy.

- Subscribe to **ImOnLinkedInNowWhat.com** to keep current on LinkedIn issues, news, thoughts, and techniques.

---

25. http://tinyurl.com/ThirstyMackay
store.harveymackay.com/SearchResults.asp?Cat=1
26. http://tinyurl.com/RICHGitomer
gitomer.com/Jeffrey-Gitomer-Little-Black-Book-of-Connections-pluL-BBC.html

Chapter 16: Complementary Tools and Resources

Conclusion

Hopefully you now have a good grasp on how to get the most out of LinkedIn, from a "how to" angle as well as a "why should this matter to me" angle.

Here are some parting thoughts:

- Know what you want to get out of LinkedIn (and related technology) and make a strategic plan to accomplish that.

- LinkedIn continually improves the product, and your experience. When I wrote this book Signals was the newest, most exciting thing, but it was still too new to write about. Focus on the goals and objectives, and use the tools, but realize what you like now might change tomorrow, and there might be some great new stuff on the horizon.

- Use LinkedIn as the tool that it is and find complementary tools to fill the gaps.

- Explore the premium upgrade options, toolbar plug-ins, and other things from LinkedIn to enhance your LinkedIn experience.

- Create a strategy to enhance your personal brand and make sure the tools and approach you use will help you execute that strategy.

- Think about career management, whether you are employed or not, happy or unhappy, a business owner or an executive. And think about how all of these tools can help you execute your career management strategy.

- Realize your entire network is not in LinkedIn. Don't neglect the rich networking opportunities outside of LinkedIn.

Do you have thoughts or ideas for the next edition of this book? Please email me: **Jason@JibberJobber.com**.

- Sign up to follow new and current topics at **ImOnLinkedInNowWhat.com**.

- If you found this book valuable, consider leaving a review on Amazon.com (hey, if Andy Sernovitz can ask, why can't I??).

LinkedIn is a great tool—but just like the power tool in your garage, it's useless until you learn how it works, and then actually put it to use!

Good luck!

"Take your connections offline. LinkedIn is a great tool, but don't forget that other tools still exist. Your phone still works, and there's nothing better than a face-to-face connection."
Scott Ingram, CEO and Founder, Network In Austin,
http://www.networkinaustin.com

"LinkedIn doesn't replace traditional networking; it facilitates it. Always supplement your online efforts with face-to-face networking."
Barbara Safani, Owner, Career Solvers,
http://www.careersolvers.com

"LinkedIn is a great tool, but you need to learn to use it and you have to maintain it to keep it sharp! Take the time to investigate all the features, encourage colleagues and friends to not only join your network but build their own so you can leverage each other's contacts, and schedule LinkedIn activities with yourself—like updating your profile once a month, answering questions daily or weekly, etc."
Deb Dib, The CEO Coach,
http://www.executivepowerbrand.com/

LinkedIn for Job Seekers

How can job seekers use LinkedIn in their job search? The same way a salesperson uses LinkedIn: finding new contacts, doing research on them, contacting them, and doing company research. Here are twelve ways a job seeker can use LinkedIn:

1. **Fix your profile.** Your profile should NOT say, "Please hire me, I'm desperate." It should say "I'm a professional, I bring value to my employer and customers." Make sure your profile is pristine, with NO typos, no grammar issues, and no noise. Craft your summary and every letter and word in your profile so you come across as professional. Remember, you are a professional in transition, not a job seeker who wants to be a professional. Right?

2. **Search for your target companies.** You should have at least five target companies you are focusing on. These are the companies you really want to work for. Search for them and find out who in your network has ever worked there, as well as what key employees you can network with.

3. **Grow your network strategically.** Find people who have contacts in your target company and invite them to connect. Work

on developing a real relationship with them. Invite them to lunch, ask for an informational interview, or call them on the phone to get more information about opportunities and other people to contact. As you add contacts from your target companies you'll get better results in your searches.

4. **Browse through your key contacts' contacts to look for professional or industry leaders with which to connect.** Ask for an Introduction and try to start a relationship with these leaders. Continue to network deeper, asking yourself, "Who do you know who...?"

5. **Use the Advanced Search page.** Search for contacts with the job titles you are applying to. Search for contacts who work, or have ever worked, at your target companies. Search for contacts that have key titles in industries you are looking for, contacts who are in certain geographic areas, etc. Find these contacts and work on contacting each of them. Look for people who can give you information you need for your job search.

6. **Ask questions regularly.** Ask questions to establish your brand with your network. Don't ask questions that can be interpreted as "I need a job, please help me!" Instead, ask questions that develop you as a subject matter expert and/or thought leader. As your network learns more about you, and what you do (or want to do), they'll be more prepared to help you.

7. **Answer questions regularly.** When you answer questions, you get the to communicate your brand with *other people's networks*. Your answers MUST be value-added, informational, and on-brand. As you answer the questions you should see new invitations to connect with people who are interested in nurturing a relationship with you.

8. **Ask for Introductions.** When you find someone you want to network your way into, or connect with on LinkedIn, use the Introductions feature. You will help your first-degree contacts understand what kinds of contacts you want to talk with, AND what your branded message is to them. Make sure you write your Introduction requests concisely, and with a compelling message, so both parties can learn more about you.

9. **Browse the Network Updates to see who you can reach out to.** You should be connected with dozens, or hundreds, of professionals. Communicate with them! Too many times I see

people not communicating with their LinkedIn Contacts, and they are missing rich opportunities to nurture relationships, communicate brands, and be in the right place at the right time.

10. **Use the job search tool to find out who else to network into.** I am not discouraging you from formally applying to posted jobs, but the real value of the Jobs section is that you can see whom you know in companies that have open opportunities and network into them. Always work to network into an open job.

11. **Find groups to join.** Look for groups that would have people in your profession, industry, target companies, or even city or state. Joining these groups should significantly increase the value you get out of Groups, as long as you actively participate in them.

12. **BONUS IDEA: Share your LinkedIn knowledge with others at job clubs and networking groups.** Become that professional in transition who shares tactics and techniques with the other job seekers (those who have the deer-in-the-headlights look). They will be forever grateful to you for sharing your knowledge, and you'll set yourself as someone who believes in givers gain. As you teach, you'll learn more about LinkedIn. This is the only suggestion of these ten that requires you to get out and talk with other people. Don't hide behind the technology!

Appendix A: LinkedIn for Job Seekers

B LinkedIn for Sales Professionals

Sales professionals can definitely use LinkedIn to increase their networks, learn about prospects, do research for sales opportunities, and communicate with decision makers. Here are eleven ideas for sales professionals to get more value out of LinkedIn:

1. **If your profile isn't complete, I might not trust who you are when you reach out to me.** Flesh it out with relevant information that helps me learn about you, and maybe even come to trust you.

2. **Search is your friend.** List keywords your target segment would have in their profile, including position/role, company, industry, interests, and associations. Go into the Advanced People Search page and do searches using advanced search strings.

3. **Grow your network shamelessly.** Okay, I mean that tongue-in-cheek, as I wouldn't tell you to blindly add people to your network. But if you want to use LinkedIn for sales, it might make sense to have a large network and be an open connector. I'm not talking hundreds of first-degree contacts; I'm talking thousands. Don't agree with me? That's okay, move on to the next point.

4. **Grow your network strategically.** Do you sell stuff in one industry? You should amass contacts from that industry. Don't worry about the person's job title. A receptionist in your target industry might have excellent contacts, don't you think? Target contacts from adjacent industries. They should have contacts in your target industry. Your growth strategy could also include certain types of professionals (accountants and CFOs, for example) or professionals in a certain geography, if your target audience is geographically based.

5. **Ask for Introductions from your connections.** When you find someone you want to communicate with, use LinkedIn Introductions to ask your first-degree contacts for a warm introduction. As you do this, you'll put your branded message in front of your first-degree contacts again. This gives you a chance to strengthen your relationship with them. Nurture individual relationships so you don't just take, take, take in the relationship.

6. **Fill in downtime when you are on the road.** Do a mile radius search to look for contacts where you'll be staying. Use keywords to narrow the results down to the right people. Consider hosting a LinkedIn get-together and inviting as many LinkedIn people as you can for dinner. Search Groups to see if there is an active local group to promote the dinner.[27]

7. **Ask questions your target audience (not your peers) would be interested in.** As they read the question, or at least your profile, they should know you are in sales, so don't try to hide that. Be genuine. Ask questions they can answer or they would want to know the answers to. You want to be known as a thought leader, and/or a subject matter expert, and/or a connector in their space. Asking questions regularly should help increase brand awareness for you and your company.

8. **Answer questions your target audience would be interested in.** Think about how others will perceive you based on your answers. Answer comprehensively, kindly, and with expertise. Share information and recommend other experts, including your customers and prospects. Use the RSS tool to learn about opportunities to chime in to new questions.

---

27. http://tinyurl.com/LinkedInBusinessTrip
jibberjobber.com/blog/2007/05/08/using-linkedin- to-fill-out-your-business-trip

9. **Join groups where your audience is or where their contacts are.** Participate in Groups Discussions. Browse through group members to look for contacts to add to your network and communicate with. Send group members messages with clear, concise messaging—focus on the relationship but let them know why you want to connect and what you have in mind.

10. **Set up the RSS feed so you get Network Updates delivered to you.** Reach out to your contacts when they have news, congratulating them on accomplishments, asking them about changes, and commenting on new connections. Use the Network Updates as an opportunity to reconnect, nurture relationships, and further brand yourself.

11. **Consider advertising on LinkedIn.** LinkedIn Advertising gives you the ability to choose certain types of LinkedIn users. It's comparatively expensive, but the ads go in front of a demographic that is supposedly above average in regard to income, professional status, and decision-making power.

About the Author

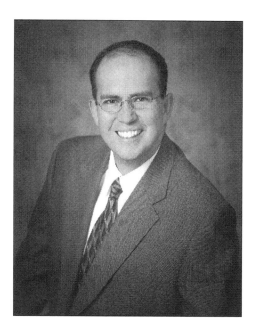

Jason Alba is *the* job seeker and networking advocate. He got laid off in January 2006, just a few weeks after Christmas. Even though he had great credentials and it was a job-seeker's market, Jason could hardly get a job interview. Finally, he decided to step back and figure out the job search process. Within a few months, he had designed a personal job search tool, Jibber-Jobber.com, which helps professionals manage career and job search activities the same way a salesperson manages prospects and customer data.

Jason is a highly rated professional speaker, sharing career management and social marketing messages to thousands of professionals each year, from California to Istanbul. Learn more at **JasonAlba.com**, or visit **LinkedInForJobSeekers.com** to get information on his LinkedIn DVD.

Other Happy About® Books

Purchase these books at Happy About http://happyabout.com or at other online and physical bookstores.

STORYTELLING ABOUT
YOUR BRAND
Online & Offline

BERNADETTE MARTIN
FOREWORD BY WILLIAM ARRUDA
AFTERWORD BY JASON ALBA

Storytelling About Your Brand Online & Offline

Bernadette Martin demonstrates how stories have transformed corporate images as well as professionals' careers.

Paperback $22.95
eBook $16.95

iPad Means Business

This book documents how the iPad is finding its place anywhere people need to be productive.

Paperback $19.95
eBook $14.95

Social Media Success!

This book is a launch pad for successful social media engagement. It shows how to identify the right networks, find the influencers, the people you want to talk to and which tools will work the best for you.

Paperback $19.95
eBook $14.95

I'm at a Networking Event—Now What???

Through this book you will learn how to make quality connections, cultivate relationships, expand your circle of influence through networking events, and create good "social capital."

Paperback $19.95
eBook $14.69

More Praise for *I'm on LinkedIn—Now What???*

"Jason promises to provide 'a solid understanding of what LinkedIn is, how to use it, and why things on LinkedIn work the way they do.' He delivers on that promise. If you're confused, frustrated, or overwhelmed by LinkedIn and wondering how to reap the benefits others have, you need the information in this book."
Winnie Anderson, Chief Strategy Maven, Virtual Marketing Mavens, http://virtualmarketingmavens.com

"Whether for job search or for making business connections, Jason Alba's book offers a logical, common sense approach to using LinkedIn. He cuts through the clutter by providing simple, concise, and remarkably well-communicated strategies for using this tool efficiently and effectively.

Far from overwhelming the reader with pages of stepwise instructions and screen prints, Jason instead shares his journey with learning and using the tool while offering still more value with additional tips and suggestions, as well as insight from other users.

I'm on LinkedIn—Now What??? *offers a quick, easy, and effective way to learn, understand, and execute a solid online strategy. It provides mindset and foundation—two important aspects to help one build more visibility and credibility. While I've used LinkedIn for years, I still found value in every chapter. A great business book and reference—and an enjoyable read—it has my complete endorsement."*
Tara Kachaturoff, Producer and Host, Michigan Entrepreneur TV, http://www.michiganentrepreneurtv.com

"Whether you are looking to advance your career or build a business, LinkedIn can be, and should be, an important part of your strategy. In I'm on LinkedIn—Now What??? *Jason Alba helps you understand how to get the most out of LinkedIn. Whether you are a LinkedIn novice or a seasoned veteran, you will find information to help take your LinkedIn experience to the next level. LinkedIn is a powerful source for job leads, career advice, sales leads, receiving invitations for speaking engagements and more. Why search for your next position when the next opportunity can find you?* I'm on LinkedIn—Now What??? *will help people with the right opportunities for you find you."*
Larry Boyer, Director of Decision Analytics, IHS, http://www.ihs.com/; **and President, Success Rockets,** http://www.successrockets.com/

"In my work as a career counselor and resume writer, Jason Alba helped me to integrate the massive confusion about how to utilize LinkedIn—as a job search tool for my clients and as a solopreneur to promote my business. There is no option for failing to learn LinkedIn if you buy this social media book."
Mindy Thomas, Founder and Principal, Thomas Career Consulting,
http://www.thomascareerconsulting.com

"Leveraging LinkedIn for your professional success takes more than publishing your professional profile and hoping someone will contact you about that next career or business opportunity. You really need to invest in learning how to get the most out of your membership. I'm On LinkedIn—Now What??? is a guidebook that will help you fast-track your success by navigating the powerful features of the platform whether you are at the start of your career or a seasoned professional."
Krishna De, Online Visibility Expert, Social Media Speaker, and Mentor,
http://www.bizgrowthnews.com

"Even though I'm a professional resume writer, I was turning down clients' requests to assist them with their LinkedIn profiles. I said I was too busy, but the truth was I was too embarrassed to admit I didn't know where to start learning about the intricacies of LinkedIn.

Jason Alba's I'm on LinkedIn—Now What??? has opened up a whole new world to me and, as a result, to my clients as well. His honest, down-to-earth, and unpretentious style makes even the most complicated aspects of LinkedIn accessible and easy to understand, including concepts like the various degrees of contacts, which had me totally confused before. His simple, easy-to-follow, and sometimes ingenious, instructions in every chapter have made it possible for me to make the best possible use of LinkedIn in my efforts to expand and market my business.

I would strongly recommend this book to anyone who, like me, signed up for a LinkedIn membership with high hopes, then found themselves baffled as to how to proceed."
Lorraine Wright, Owner, 21st Century Resumes,
http://www.21stcenturyresumes.ca

"Alba's book is the link you need to power up your LinkedIn results!"
Kent M. Blumberg, Executive and Professional Coach, Kent Blumberg Partners, http://www.KentBlumberg.com

"A few years ago, Jason Alba established his well-deserved place in the careers field as the go-to expert for LinkedIn. We look to him for advice on how to maximize LinkedIn for clients as well as seek his guidance about using LinkedIn to promote our businesses that cater to the millions of LinkedIn members. With each revision, he has added new and valuable tips. His latest edition is up-to-date, practical, and insightful. Jason rules LinkedIn!"
Debra Feldman, Executive Talent Agent, JobWhiz,
http://www.jobwhiz.com

"I was ready to abandon my LinkedIn account before I read Jason Alba's concise and remarkably useful guide. Jason writes with remarkable clarity, provides one useful tip after another about how to use it most effectively, and unlike so many users guides that offer breathless and uncritical hype, Jason candidly explains the virtues and drawbacks of LinkedIn's features. Beyond that, Jason has such deep experience with the Web that the book contains hundreds of broader lessons about how to get the most of the Web. I learned an enormous amount from this little gem"
Robert Sutton, Professor, Stanford University; and Author, *The No Asshole Rule*, http://bobsutton.typepad.com/

"Jason offers a unique perspective on networking that's of interest to anyone that is a job seeker, entrepreneur, or networking enthusiast. He has been all of these and his experiences with LinkedIn enable him to offer an integrated review for anyone to make the most of the LinkedIn tool. His book is a reflection of his deep understanding of people, technology and change in the market and can easily save the average new user months of time in trial and error."
Nadine Turner, Sarbanes-Oxley (SOX) IT Program Manager; and Six Sigma Black Belt

"If you are new to LinkedIn, you are in for a treat when you read I'm On LinkedIn—Now What??? *If this book were available the first year LinkedIn started, it would have helped LinkedIn to be better understood and would have helped thousands of professionals get the most out of LinkedIn."*
Vincent Wright, Chief Encouragement Officer, http://www.brandergy.com

"In an age of social networking, LinkedIn remains one of the best for business people. Jason gives a wonderful firsthand insight on the how-to's of using the service: this guide has been a long time coming. I am delighted that he's taken the time to put together, in a single volume, how to get the best out of the service."
Jack Yan, CEO, Jack Yan & Associates, http://www.jackyan.com

"Jason has written a great book I'm on LinkedIn—Now What??? for the beginner and those that need some more detailed instruction. This book is an easy read with some great descriptions of how to accomplish your LinkedIn networking tasks. I recommend this book for all users of LinkedIn."
Jim Browning, Co-owner/Lead Moderator, LinkedIn Atlanta; and President, Browning Business Solutions, LLC, http://www.networkingga.com

"Jason has written a highly practical guide to LinkedIn that will quickly allow a new user to understand and utilize LinkedIn. It's also a great guide to the LinkedIn's hidden gems—finding high quality people through endorsements and off LinkedIn content such as Groups and identifying thought leaders through blogs linked from profiles."
David Dalka, Senior Marketing and Business Development Professional, http://www.daviddalka.com

"This is a great book. I think Jason did an excellent job. I would recommend it to anyone who is just starting to build a LinkedIn network or for someone who has been a member for some time but is just now seeing the advantages LinkedIn provides."
Thom Allen, Writer and WordPress Consultant, http://www.thomallen.com/

"I'm on LinkedIn—Now What??? provides a useful guide for all those looking to better utilize the power of LinkedIn. As Jason writes, LinkedIn is NOT the silver bullet of networking sites; such a site does not exist, and this book does not try to make that point. What this book does incredibly well is show how you CAN use the tool to your advantage—to make connections, to help others, and ultimately, to help yourself! Two handshakes WAY UP for this great book!"
Phil Gerbyshak, Author, *10 Ways to Make It Great!*; Owner, The Make It Great Guy, http://www.makeitgreatguy.com

"Jason's book is an easy-to-read, well-written, step-by-step tutorial for the novice, or for the person who's already linked in. He reveals his mastery, once again, at making the complex simple, just as he did with his invention of JibberJobber."
Billie R. Sucher, Career Transition Consultant; and Author, *Happy About the Career Alphabet*, http://www.billiesucher.com

"Jason Alba has established himself as a well-known and widely respected expert in the employment arena. His success in establishing himself and promoting his extraordinary career toolset JibberJobber.com prove that he knows what he's talking about. His understanding of personal branding and networking come together in his new book about using LinkedIn. Authoritative and insightful, this book is a great primer for 'newbies,' yet it's comprehensive enough to offer something of value to even the most seasoned LinkedIn users."
George Blomgren, Talent Acquisition Strategist

"You don't have to be a full-time social networker to use LinkedIn as a connection making tool and Jason Alba lays this out point by point in I'm on LinkedIn—Now What??? *in which he explains exactly how to use LinkedIn in a way that works for you. It's a book that I've long needed to explain just what LinkedIn is and isn't to countless friends and clients without buzzwords or hype; it's high on my list of recommendations."*
Susan Reynolds, New Media Consultant, http://www.artsyasylum.com

"As more and more business professionals hear about LinkedIn, they're looking for a place to go for answers about how to get involved and effectively use this important tool. Jason's book—I'm on LinkedIn—Now What??? is appropriately titled and is the quickest and easiest way to understand what LinkedIn is, it's purpose, and how to effectively use it. I'm an avid LinkedIn user and regularly teach classes to busy professionals on how to use it for networking, job search, business development, or recruiting. One of the first things I do in all of my sessions is to recommend the book and the companion blog, because there are no other resources out there that cover how to get started and how to effectively use LinkedIn as well!"
Jennifer McClure, Executive Recruiter/Executive Coach; and President, Unbridled Talent, LLC, http://unbridledtalent.com/blog

"Jason has a great knack of explaining the features and benefits of LinkedIn in a way that doesn't intimidate a novice. Yet he also includes little gems that can benefit even seasoned LinkedIn users. He clearly demonstrates how readers can benefit from LinkedIn in every chapter of I'm on LinkedIn—Now What??? *Readers will understand why they should do something rather than just being told that they should do it. This will bring more value to their LinkedIn experience."*
Christine Pilch, Co-owner of Grow My Company and Coauthor, *Understanding Brand Strategies: The Professional Service Firm's Guide to Growth,* http://growmyco.com

"*Jason's Personal Brand is consistent in each project he works on, especially in I'm on LinkedIn—Now What??? Throughout this book he narrows down exactly what LinkedIn SHOULD be used for so that readers don't confuse it with other social networks. You will encounter information on how to set up your profile, network through Groups and proper etiquette to use as you grow your LinkedIn database. Jason's thoughtful and honest viewpoint on LinkedIn will teach everyone from youthful professionals to experienced entrepreneurs how to succeed with this tool.*"
Dan Schawbel, Managing Partner, Millennial Branding LLC; Author, *Me 2.0*; and Publisher, Personal Branding Blog,
http://www.personalbrandingblog.com

Made in the USA
Lexington, KY
05 March 2013